Ozone and Climate Change

Ozone and Climate Change

A Beginner's Guide

Stephen J. Reid

National Oceanic and Atmospheric Administration
Colorado, USA

Gordon and Breach Science Publishers
Australia • Canada • France • Germany • India • Japan
Luxembourg • Malaysia • The Netherlands • Russia
Singapore • Switzerland

Amsteldijk 166
1st Floor
1079 LH Amsterdam
The Netherlands

British Library Cataloguing in Publication Data

A catalogue record for this book is available from the British Library.

ISBN: 90-5699-232-5 (hc)

To Ria, who encouraged me to write this book,

Richard, who *almost* read it,

and Rog, without whose help

it would have been finished in half the time.

CONTENTS

INTRODUCTION

Since the beginning of the industrial revolution early in the nineteenth century, our ability to change the world around us has become profound. At first, the impact on our planet was almost imperceptible, but as we have grown both in number and technological capability, that influence has grown with us. Lately, the effects of our increased activity have begun to manifest themselves in a multitude of subtle, and some not so subtle, ways: we have thinned the ozone layer and may now be starting to change the very climate system upon which we and all other life on Earth depend. In effect, we are experimenting with the future, but unlike performing a laboratory experiment which can be scrapped and begun anew if its fails, altering the climate is something that cannot easily be undone. Whatever happens, we shall all be forced to live with the consequences for a very long time.

But isn't this all rather extreme? Indeed, is there really a serious problem at all? So far, the Earth seems to be capable of dealing with whatever we throw at it. Should we be concerned? Certainly, there is now an opinion abroad that the ozone problem has been solved, primarily because the chemicals which created it are no longer produced. There is also heated controversy as to whether or not climate change is even an issue. Some believe it is nothing more than the hype of an alarmist minority whose speculations may ultimately be harmful to the economy. Is this true?

It is certainly difficult to make sound judgements about such matters when we are not presented with all the facts. Those we do hear through the media and from politicians are often quoted out of context to accentuate only one aspect of the problem, either to make a better story or to promote a particular agenda. We often hear the phrases *the ozone hole* or *the greenhouse effect*, but what do they *really* mean? Does the former conjure up the impression of a hole spanning the entire depth of the atmosphere, and the latter a world made hot and arid by greenhouse gases? In both cases, this is only a partial truth.

Matters become still more confusing when one hears that members of the scientific community specialising in these areas of research cannot agree amongst themselves. Even today, there are still a handful of scientists who insist that the ozone hole over Antarctica is a naturally occurring event, flying in the face of overwhelming evidence to the contrary. Given that the ozone depletion problem is actually quite straightforward, if considered in the light of the stunning complexity of the climate as a whole, what hope is there that they will agree on its future? This is not merely an academic problem: actions taken to safeguard the environment of the future are likely to negatively affect economic industrial growth, especially in the developing countries of the world. But how can we expect governments to make appropriate policy when they receive conflicting views from different quarters? After all, if the scientific community cannot agree, how can the public be expected to arrive at a concensus?

Fortunately, the situation is not as bad as it seems. All that is required is that we fill in the gaps in our knowledge so that we can make sound judgements, at least as far as the limits of that knowledge will allow. The first misconception we should confront is the tendency to make sweeping assumptions about climate change based on incomplete evidence. Most of us have an opinion about the climate; for the British (according to international opinion) it is an all-consuming preoccupation! It is certainly tempting to believe our own senses and deduce that the environment in which we live is indeed changing. The extremes in the weather over the past several decades, during which most meteorological records have been repeatedly broken, do seem to support this belief. Consider, for example, the extraordinary flooding throughout many parts of Britain in recent years, not the mention the devastating hurricane which cut a swathe through the south-eastern counties on 16th October 1987. The United States has also experienced its fair share of apparently abnormal weather lately: on 22nd September 1989, Hurricane Hugo swept across South Carolina with wind speeds in excess of 200 kilometres-per-hour (km h^{-1}), locally lifting the Atlantic ocean by almost seven metres and driving waves which were higher still; Hurricane Andrew, which came ashore in south Florida on 24th August 1992 with winds exceeding 260 km h^{-1}, caused more than \$20 billion (approximately £13 billion) worth of damage, killing thirty people and leaving a quarter of a million more homeless.

In Africa, the annual rains have frequently failed over the past few decades, leaving the continent in the grip of an unrelenting drought, whilst early in 1996 Scandinavia experienced inland temperatures of –50°C, comparable to those normally found on the Siberian plateau. Across the world in Australia and Indonesia, drought induced by the 1997 El Niño warming event in the equatorial Pacific resulted in bush fires which raged for months.

There seems to be evidence too that the world is warmer and drier than it used to be. In July 1976, England experienced the hottest month in the past three hundred years, whilst June 1991, in complete contrast, was one of the dullest, coolest and wettest months of the twentieth century, despite an overall *decrease* in the average annual rainfall in the U.K. throughout the preceding decade. Such examples of local extremes go on *ad infinitum*, perhaps creating the impression that what might once have been considered exceptional weather is fast becoming the rule.

Or is it simply that media coverage is now far more extensive than in the past, keeping us better informed even about conditions in parts of the world that, a few decades ago, rarely figured in the news? Are unusual weather phenomena really confined to recent times? Looking back through meteorological records, one finds that 1947 boasted the wettest March in 250 years, whilst the lowest temperature recorded in Britain, –27.2°C, occurred in the small Scottish town of Braemer in 1895. But these dates still lie within the span of the industrial revolution, so might they not still have arisen from human activities? During the late 17th and early 18th centuries, decades before the industrial revolution, the winters in the northern hemisphere were unusually cold for more than seventy years, and in a particularly cool spell lasting from January to March 1709, temperatures were low enough for the River Thames to freeze over completely. Clearly, making *ad-hoc* predictions about future climate based on isolated observations of weather extremes is a highly dubious practice.

To make any real progress in understanding the true nature of the climate, we must first understand how the atmosphere, the cradle of the climate system, works. This is where our weakness lies. Many misconceptions stem from an incomplete knowledge of the underlying principles connected with the atmosphere, without which it is difficult to understand why ozone loss is most severe over the Antarctic when the chemicals which give rise to it are primarily released in the northern hemisphere, or why it is that a warming event in the equatorial Pacific (El Niño) can induce drought and widespread fires in Indonesia and Australia which are far removed from the site of the action.

To rectify this problem, the first part of this book will guide you through some of the fundamental principles about the atmosphere which will help you to understand how the atmosphere works, at least in a general sense. These chapters are not intended as a comprehensive introduction to meteorology, and the interested reader should consult the bibliography for this section of the book for sources of further reading.

Scientists invariably evolve a considerable amount of terminology to describe the phenomena they study in their various disciplines, and the atmospheric scientific community is no exception. A book which ignores this terminology completely, whilst accessible to more people, will of necessity omit a great deal of important information. In an effort to address both needs, this book has been designed so that the main body of the text contains no mathematics at all opting instead for verbal explanations. For the reader who wishes to explore the subject a little more formally, a set of appendices has been provided at the end of each section which require a knowledge of advanced school/college level mathematics, including some elementary calculus. The sole exception to this rule is the inclusion of some basic chemistry in Part Two when discussing how ozone is formed and destroyed both naturally and anthropogenically. This subject is explored in greater depth in Appendix 2.4. The Periodic Table has also been included for completeness, since it demonstrates how man-made ozone destruction arises essentially from a single group of elements.

Part One
The Earth's Atmosphere

CHAPTER 1.1

A QUESTION OF BALANCE

The atmosphere around us is in a perpetual state of motion, and it is hard to imagine that any kind of order may be imposed upon it. We are still unable to generate a 24-hour weather forecast with any real certainty, still less anticipate when or where a gust of wind may occur. In fact, the atmosphere is inherently chaotic on these small-scales, and it is essentially these which impair weather forecasting.

But given this inherent unpredictability, why is it that weather forecasting works at all? It does so only because on the scale of weather systems (hundreds to thousands of kilometres), the atmosphere actually *is* reasonably predictable. (In this book, the very large and very small scales will crop up fairly frequently, and Appendix 1.1 explains the use of the exponent form employed to represent large and small quantities).

The large-scale predictability of the atmosphere is our saving grace when trying to follow the transport of the air, and the chemistry it supports, over a period of time. This too may seem an unlikely concept: the idea that air can be tagged in some way and followed around the globe may seem improbable, but to some extent it can be done; the concept of a *parcel of air* is not merely an abstraction employed by theoretical physicists. Making use of well established rules which govern nature, including Newton's First Law, we have discovered how chlorofluorocarbons (CFCs) are able to travel into the upper atmosphere (despite being heavier than air) where they break apart; and we can follow the progress of their reactive components as they travel to the Antarctic and Arctic where they set about reacting with, and destroying, ozone. These rules also enable us to understand how a warming event in one part of the world, El Niño, can affect places far removed from its source.

First, however we need to put aside our everyday experiences of clouds and precipitation, and wind gusts and short-lived fluctuations in temperature, since these all belong to the small-scale regime of atmospheric motion. On the broader canvas, there are just a few major players which control the atmosphere, keeping it in a dynamic state of balance:

(i) atmospheric pressure (in the vertical and horizontal)
(ii) the downward acceleration of gravity
(iii) the Coriolis Effect, produced by the Earth's rotation

The last of these is perhaps the least obvious of all forces, because we rarely think about living on a rotating planet, and we tend instead to ascribe its effects to some imaginary force. Whilst not a force in itself, the other major player in the atmosphere is temperature, to which the speed and direction of the wind is intimately linked.

Let us begin by examining the forces which control the atmosphere in the vertical direction. Going from the surface of the Earth to the edge of interplanetary space, the atmosphere is classified in terms of temperature.

(i)   The vertical forces of the atmosphere

There are four distinctly different temperature regimes in the atmosphere: the *troposphere, stratosphere, mesosphere* and *thermosphere.* Figure 1 is a schematic which shows how temperature varies with altitude.

The lowest part of the atmosphere, between the ground and roughly 8 km deep at high latitudes, 10–12 km deep at mid-latitudes and 15 km deep near the equator, is called the *troposphere*, from the Greek word *tropos*, meaning *change* (or, more loosely, *overturning*). This is the region of the atmosphere where practically all life exists (neglecting the oceans). Temperature drops sharply with increasing altitude throughout the depth of the troposphere, falling at a more or less constant rate of 7°C for each kilometre increase in height (or about 15°F for each mile). Because temperature drops so swiftly in the troposphere, the air is very unstable in the vertical, allowing it to rise and fall and over-turn rapidly (hence the name troposphere). This rapid vertical motion, called convection, mixes together all of the gases in this part of the atmosphere very effectively, so that any gases released by industry, for example, will quickly find their way up to great heights. In fact, their vertical progress is impeded only by the barrier between the troposphere and the region above it. This barrier, where the temperature stops falling with height, is called the *tropopause.*

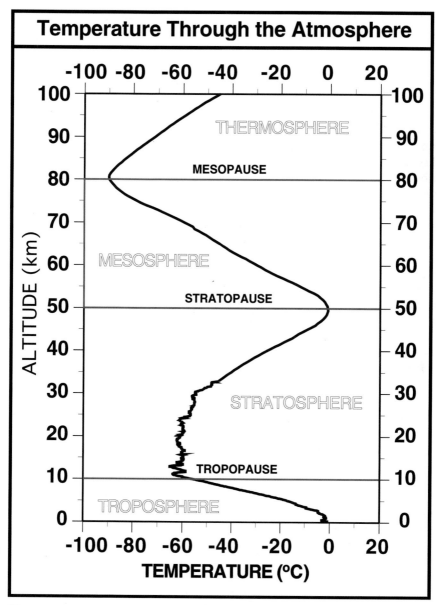

Figure 1 Temperature from ground level to the top of the atmosphere, in degrees Celsius. The atmosphere is divided into regions according to whether temperature is falling with increasing height (troposphere and mesosphere) or rising with increasing height (stratosphere and thermosphere).

The first region where the temperature remains constant or increases with height is called the *stratosphere: stratos* is the Latin word meaning *layer*. The stratosphere extends from the tropopause up to around 50 km. The reason the temperature has now stopped falling with height is due to the presence of the ozone layer. It is here, above the region where most aircraft fly (at least at the moment) that nearly all atmospheric ozone is naturally produced and destroyed (ignoring the tiny amounts generated by the ocean and lightning storms near ground level). But why should the ozone layer affect the temperature? It is because ozone very efficiently absorbs energy from the sun over a narrow band of wavelengths, that part of the electromagnetic spectrum (see Appendix 1.4 for an explanation) which contains ultraviolet radiation. The subject of ozone and UV radiation is discussed in chapter 2.1, but for now it is sufficient to appreciate that the absorption of UV radiation deposits a considerable amount of heat energy into the atmosphere, warming it over a vertical depth of 35–40 km. For a gas that is extremely rare (typically present only in concentrations of around five molecules in every million molecules of air), ozone has a profound influence on temperature. Viewed from a different perspective, the ozone layer is actually responsible for the existence of the stratosphere.

The increase in temperature with height makes the stratosphere a very stable place where air does not (generally speaking) overturn in the vertical. In contrast with the troposphere, where vertical wind speeds are often several metres per second, in the stratosphere they are seldom more than a few millimetres (or a few centimetres) per second. Consequently, it takes air a very long time to be transferred from the bottom to the top of the stratosphere; years in fact. This inability to mix rapidly in the vertical direction is the principle reason why CFCs, the substances which carry ozone-destroying chemicals into the stratosphere (see chapter 2.2), take so long to reach altitudes where the Sun's energy is sufficient to break them apart, and why some of them will still be there a hundred years (or more) from now.

At the top of the stratosphere, where there is little ozone to absorb energy and heat the atmosphere, there is another temperature inversion called the *stratopause*, above which lies the *mesosphere*, the region between 50 and 80 km where temperature again falls with height. The *mesopause* is located at around 80 km where temperature again stops decreasing. Above this level, the *thermosphere* stretches up to the frontier of interplanetary space. At these high altitudes, the atmosphere is bathed in short-wave radiation from the Sun. Even visible light is now sufficiently energetic to strip electrons away from their orbits around atomic nuclei, leaving behind electrically-charged particles known as *plasma*. In other words, this region of the atmosphere is completely *ionised*.

Neither the mesosphere or thermosphere are germane to the focus of this book, and they will not be discussed further. However, for those interested in the upper atmosphere, references to sources of more information are available in the end-of-section bibliography.

Another thing that happens as altitude above the Earth's surface increases is that atmosphere pressure decreases. Witness, for example, the difficulties experienced by mountain climbers on Mount Everest, in the Himalayas, whose summit stands at 8,850 metres (or slightly more than 29,000 feet) above sea level. On approaching the summit, climbers are forced to breathe oxygen from cylinders which they must carry with them. Nor is it only the number of oxygen molecules which diminishes with increasing height; *all* the gases which make up the atmosphere become increasingly sparse. As the number of molecules in a given volume decreases, so the pressure they exert by colliding with one another also decreases as the collisions become less frequent. At the top of Everest, the pressure exerted by the atmosphere is only 28% of that at sea level, causing water to boil at around 30°C. (The relationship between pressure and temperature, the *Ideal Gas Law*, is explained in Appendix 1.2).

As indicated by figure 1, temperature also decreases rapidly with height in the first 10–15 km above the Earth's surface, seldom rising above –40°C among the Himalayan peaks.

The decreasing pressure of the atmosphere with height means that, given a choice, the air around us would rapidly leak away into space as it moved down the pressure gradient to the lowest possible energy state, something that nature is always trying to achieve. In a simplistic sense, pressure tries to move the air upwards as water always flows to the lowest point on the ground. The force controlling them both is the same: gravity. The pull of gravity towards the centre of the Earth exactly balances the pressure gradient force acting upwards, creating a state of balance which serves to hold the atmosphere in place. A simple mathematical treatment is provided in Appendix 1.2.

(ii) The horizontal forces in the atmosphere

Whilst gravity inhibits the upward flow of the atmosphere, there is no obvious analogy in the horizontal direction, no tangible force which prevents wind speeds from becoming infinitely large. We are not really aware of the horizontal restraint because it lies outside of our everyday experience; it is an imaginary force which is not really a force at all. It is called the *Coriolis Effect*, and is imaginary because it is an effect induced by the rotation of the Earth. For the sake of completeness, the Coriolis Effect also has a vertical component, but this is sufficiently small that we may ignore it.

The Coriolis Effect it not properly incorporated into the Newtonian equations of motions, and to compensate for its presence, an extra term has been added to the equations in order that we may continue pretending that we are living on a stationary world (or an inertial frame of reference, implicitly assumed in Newton's equations), instead of none which is rotating. The Coriolis Effect limits horizontal differences in pressure in the atmosphere, preventing them from becoming infinitely large and therefore placing a cap on wind speeds (see Appendix 1.3).

There is not only a gradient in pressure acting upwards in the atmosphere, but also another acting in the horizontal direction, albeit very much smaller in magnitude. This pressure gradient is primarily due to differences in temperature arising from the asymmetric distribution of sunlight with latitude. Up to an altitude of around 5 km, the Earth receives most of its energy from the Sun in the tropics, and least in the polar regions. This drives surfaces of constant pressure (isobars) farther apart at the equator than at the pole where it is colder. Thus, the isobars slope down as latitude increases, creating a horizontal gradient in pressure. Air at high latitudes (seeking the lowest state of energy) flows towards the region of lower pressure nearer the equator. In other words, a wind flows from pole to equator.

At heights greater than around 5 km, the situation is reversed; the atmosphere is now warmer at higher latitudes and pressure surfaces slope downward towards the equator, creating a wind blowing *towards* the pole where the pressure is now lower. Fortunately, it never completes this journey. If it did, wind strengths in the atmosphere would be far stronger than they are. Its journey is halted by the Coriolis Effect.

Because neither the atmosphere nor the oceans are rigidly bound to the Earth, they both *feel* the influence of the Coriolis

Effect which turns air (and water) in the northern hemisphere to the right of its direction of flow. Moreover, the Coriolis Effect is not uniform everywhere on the Earth. Its influence depends on wind speed and latitude; it is zero at the equator and reaches a maximum at the geographic pole. Thus, as an air parcel moves away from the equator, propelled by a gradient in pressure it appears to undergo a steadily increasing deflection in the direction of the Earth rotation, which acts to turn it clockwise in the northern hemisphere and counter-clockwise in the southern hemisphere. Eventually, the Coriolis Effect is strong enough to overcome the pressure gradient force and the air turns back towards the equator. As it moves away from the pole, however, it starts to slow down as the Coriolis Effect weakens and the pressure gradient force increases again until, at some point, the latter overcomes the former and the air turns polewards once more. This tug of war continues as the air circumnavigates the entire hemisphere. The oscillating balance between the horizontal pressure gradient force and the Coriolis Effect is shown schematically in figure 2.

This type of flow of air is given a special name in meteorology; it is called *geostrophic* flow, or the *geostrophic wind*.

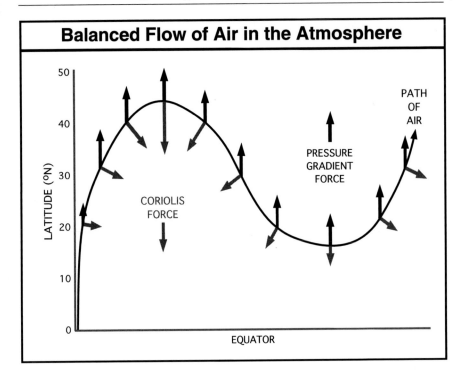

Figure 2 A question of balance: air in the troposphere flows poleward in response to a horizontal gradient in pressure, comparable to water flowing down-hill under the force of gravity. As it moves away from the equator, the air starts to feel the Coriolis Effect which turns it to the right of its direction of flow (in the northern hemisphere), eventually steering it back towards the equator. These two forces are approximately balanced, making the air flow around the hemisphere in a wave-like motion called *geostrophic flow*.

THE ZONAL AND MERIDIONAL CIRCULATIONS

The flow of air around the Earth is not everywhere the same. There are regions where air moves rapidly relative to the ground, travelling eastward in narrow bands at speeds of between 40 and 100 ms^{-1}. These bands of high-velocity wind are called jet streams, named so by the meteorologist Carl-Gustav Rossby in 1947. In every important respect, they control the weather over the entire planet. For example, an El Niño warming in the tropical Pacific can change the position of a jet stream located at the top of the subtropical troposphere, triggering severe storms and droughts in places remote from the source of the disturbance (chapter 3.3). A second tropospheric jet stream is typically found between 40° and 60° north (and south), whilst in the winter stratosphere, another powerful jet stream, called the *polar night jet*, circles the polar regions, confining a large volume of air at high latitudes. Without this last, there would be very little man-made ozone depletion. The positions of the northern hemisphere's major jet streams are shown in the schematic in figure 3. Their distribution is similar in the southern hemisphere.

How do jet streams form? There are two principal contributors which lead to their formation: (a) the distance from the axis of rotation of the Earth to the surface changes with latitude, it is greatest at the equator and diminishes on moving to higher latitudes, falling to zero at the geographic pole; (b) Newton's First Law, which states: *an object remains in a condition of rest or of uniform motion in a straight line unless acted upon by a net external force.*

The Sub-Tropical Tropospheric Jet Streams

Imagine standing on the equator, and the air around us is still; that is, there is no wind. Why? The ground at the equator is moving in a large circle at a speed of 1,670 kilometres every hour, or 463 ms^{-1}. The air around us seems still, but only because it too

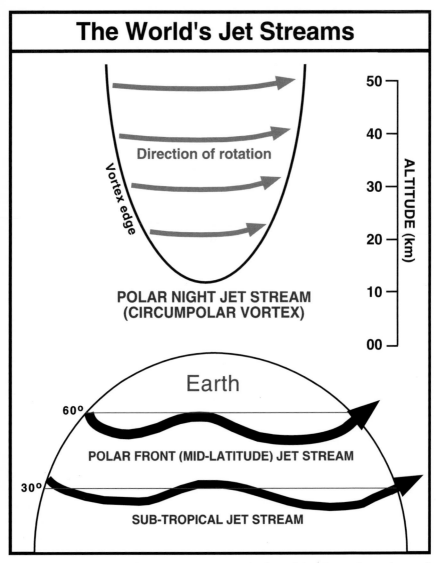

Figure 3 The world's jet streams: one is found in the subtropics and one in mid-latitudes, both near the top of the troposphere (around 10–15 km above the ground). Another jet stream appears in the polar stratosphere during winter, forming a semi-permeable containment vessel inside which ozone is destroyed by man-made chemicals.

is moving at 463 ms^{-1}. The atmosphere, not being rigidly attached to the Earth, is free to move relative to the ground. At the equator, incoming heat from the Sun warms the tropical atmosphere and the air expands and rises (it is now less dense than the air above it, like wood rising to the surface in water). This heating produces large and rapid vertical motions throughout the depth of the tropical troposphere (about 15 km). As the air cools aloft, the water vapour it carries condenses to form clouds, giving rise to the strong convective cells responsible for tropical thunderstorms. When this air reaches the tropopause, the boundary between troposphere and stratosphere, most of it stops climbing and flows horizontally poleward, down the temperature, and hence pressure, gradients in both hemispheres. A certain amount of the air overshoots the tropopause, however, pushing its way through into the strato-sphere above. It is this overshooting process which conveys large volumes of tropospheric air into the stratosphere every day.

The air moving to higher latitudes at the top of the troposphere is now travelling eastward with the same speed it possessed at the equator, but now the Earth's surface is turning more slowly beneath it: recall that the distance of the Earth's surface from its axis of rotation decreases as latitude increases. By the time the air from the equator has moved to 10°N, it is already travelling 7 ms^{-1} faster than the ground; at 20° latitude, 28 ms^{-1} faster, and at 30° latitude, 60 ms^{-1} faster. The air is turning towards the right of its direction of motion with steadily increasing speed. But for the increasing magnitude of this deflection with latitude, the air would flow all the way from the equator to the pole where, un-checked, it would form an infinitely large and permanent tornado (a very powerful vortex) that would rage throughout the year. It is hard to imagine life evolving on a planet sheathed in such ferocious winds.

The poleward flow of air is halted in the subtropics simply because the Earth is spinning too fast. This may seem contrary to common sense; the faster the Earth spins, the stronger the Coriolis force becomes and the greater the tendency for the atmos-phere to spin up into vortices. However, experiments have revealed that above a certain speed of rotation, the motion of the atmos-phere breaks down into a pattern of symmetrical waves, disrupting the equator-to-pole motion. These waves are produced in part by the balance of forces described in the previous chapter, inducing oscillations in the atmospheric flow as the pressure gradient force and Coriolis Effect vie for dominance. Waves are also formed when the atmosphere flows over mountain ranges, such as the Himalayas or the Rockies. In this situation, air initially at sea level is forced to rise several kilometres in height in order to flow over the mountains, setting up oscillatory motions in the vertical. In fact, the atmosphere is full of wave-like motions which do much to disturb its flow around the globe, especially in the northern hemisphere.

The subtropical jet stream exists, then, because at 30°N the Earth's rotation halts the poleward flow of air and large, horizontal circulations take over. Forced downwards (unable to rise higher into the stable realm of the stratosphere), the air sinks and warms. Clouds evaporate because of the heating, giving rise to the subtropical band of fair weather (high pressure) where very little rain falls — witness the Sahara and the Indian continent. As it nears the ground, the wind speed slows down and even reverses direction, now flowing weakly from east to west, creating the so-called Trade Winds of subtropical latitudes. A dragging force (friction) acts on the air as it moves near the Earth's surface, slowing it down dramatically and preventing the Trade Winds from possessing jet-stream velocities. This circulation pattern in the tropics is known as the Hadley Circulation, sometimes the Hadley Cell.

During the northern hemisphere summer, another jet stream forms in the tropics, this one flowing in an easterly (from the east) direction. It is found between 10° and 20°N, usually a little to the south of high land masses in Asia and Africa. At this time of year, there is warm air to the north and cold air to the south (the opposite of the wintertime situation), which means there is a gradient in temperature, which in turn creates what is generally known as the Tropical Easterly Jet, although in fact it is located more than a thousand kilometres from the equator.

The Mid-Latitude Tropospheric Jet Stream (The Polar Front Jet)

The poleward transport of tropical air does not entirely stop in the subtropics, but it is halted by the presence of a second tropospheric jet stream which is generally found between 50°N and 60°N during summer when it moves eastward at 25–50 ms^{-1}, and 35°N to 45°N in winter when its speed increases to around 50–75 ms^{-1}: this is the Polar Front Jet. This jet stream, approximately 150 km in north-south (meridional) extent and only 3 km deep, is a geostrophic wind created by the large pressure gradient between warm tropical air and cold polar air, and the Coriolis Effect. It is visible on meteorological charts where lines of constant temperature, called isotherms, are most tightly packed. It is not a continuous jet, however, and is frequently distorted by developing weather systems which cause it to disappear altogether over considerable distances. The jet later reforms, but often at a different altitude.

When weather forecasters talk about a jet stream, they are generally referring to the Polar Front Jet. As indicated above, it lies at the boundary of polar and subtropical air masses; on the polar side, temperature, and therefore atmospheric pressure, decrease more rapidly with height than in the warmer air mass to the south, which sets up a pressure gradient and hence this very strong belt of winds. Indeed, temperature is critical in controlling the location and strength of this and the other jet streams, as you will see in Part Three in connection with El Niño. A small change in their velocity or direction produces strong vertical motions in the air below them, which plays a key role in the development of high and low pressure weather systems.

The Stratospheric (Polar Night) Jet Stream

Whilst the sub-tropical and mid-latitude jet streams are both found in the troposphere, there is a third, very powerful jet stream which appears in the high-latitude stratosphere during early winter, between ~15 and 50 kilometres above the surface of the Earth. This jet forms in the autumn and persists throughout the winter, creating a spinning vortex over the pole which breaks down only when the sun rises over the Arctic (or Antarctic) in spring (although in the northern hemisphere there are sometimes sudden warming events during the winter months which can weaken or even dissipate this jet). The boundary of this stratospheric polar vortex, where the wind speeds are highest, is generally located close to the polar circle, the line demarcating night and day during the polar winter. Incursions of large, warm air masses from further equatorward can displace the vortex from the polar regions, however, and one or twice in each winter it will extend over Canada, Europe and Russia.

The horizontal area covered by the polar vortex is greater at the top than at the bottom, so that it is shaped somewhat like a giant ice-cream cone, with the vertex of the cone cut away. This cone-like structure acts like a semi-porous containment vessel, inside which ozone-depleting chemistry can proceed with startling efficiency (chapter 2.2).

Throughout the entire winter and often well into spring, most of the air caught up in this containment vessel remains trapped there, interacting little with the rest of the atmosphere. As the winter proceeds, more and more air piles in at the top of the cone via an equator-to-pole circulation (the Meridional Circulati n; see later in this chapter) and sinks down into the cold polar night, carrying with it more ozone as well as the chemicals which will bring about ozone destruction. Some air does leak out from the bottom and from the sides of the cone, often in the form of long filaments which snake away into mid-latitudes, but this exit route probably only removes a small percentage of the total mass of air every month.

Over the Antarctic, this huge spinning vortex of air maintains a far higher degree of integrity than its Arctic counterpart, because in the southern hemisphere there are no mountains to disrupt its vortical motion. In the northern hemisphere, the Rocky Mountains, the Greenland Plateau, and the Urals and Verkhoyanskiy ranges in Russia all serve to deflect the flow of the atmosphere. As a result, standing waves (sometimes called mountain waves) are set up and their influence is felt far into the stratosphere. These standing waves, combined with the far bigger and mobile planetary-scale waves which propagate through much of the atmosphere, disrupt the Arctic polar vortex almost every winter, sometimes more than once, permitting the large-scale ingress and egress of air which keeps the whole system warmer. In consequence, the Arctic vortex not an efficient site for ozone destruction, although this may be changing (chapter 2.3).

In addition to the prevailing motion in the direction of the Earth's rotation, there is another, more subtle circulation which carries air from the tropics to the polar regions: the Meridional Circulation, also known as the General Circulation. This is the residual movement of the atmosphere after all other forces have been accounted for, and it is driven by the temperature difference between the equator and the pole. For this reason, its flow is strongest in the winter hemisphere.

As the schematic in figure 4 shows, air rises at the equator, lifted by powerful convection, and quickly traverses the depth of the troposphere (capped by the tropopause, the thick black curve running from around 8–10 km at high latitudes and rising to

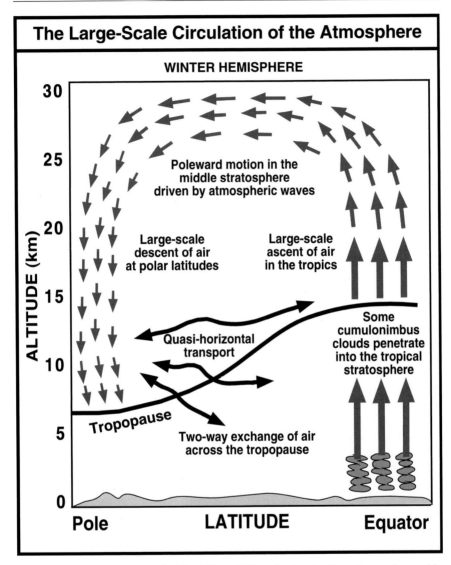

Figure 4 The large-scale Meridional Circulation in the atmosphere. Air at the equator is lifted by deep, convective thunderstorms over the tropics, reaching far into the stratosphere. Most of the air then flows into the winter hemisphere and sinks over the polar regions before returning to the troposphere across the tropopause, the temperature barrier which divides the troposphere from the stratosphere.

around 15 km in the tropics). As mentioned when discussing the Hadley Circulation, most of this air is deflected poleward where it later sinks back to the Earth's surface, but some, carried by its own momentum, overshoots the tropopause and enters the stratosphere. Once there, it continues rising, albeit slowly now, to still greater heights.

In the middle and upper part of the stratosphere, this flow turns poleward in response to the temperature gradient, cooling and becoming denser as it travels into higher latitudes. As its density increases it begins descending, either into the wintertime polar vortex to build up large concentrations of ozone (and CFCs), or just outside the vortex, allowing ozone to accumulate to form a collar around its periphery. At the bottom and the sides of the vortex, the air seeps out and returns to lower latitudes, re-entering the troposphere via exchange processes which carry air back and forth across the tropopause. This process accelerates dramatically in spring when the polar vortex is broken down by the return of the Sun to polar latitudes.

It is the Meridional Circulation which carries man-made pollutants (the vast majority of which are released in the northern hemisphere mid-latitudes) from the equatorial regions to the polar stratosphere in both hemispheres. Whilst slow, it is sufficiently powerful to overcome the tendency of molecules to settle out by weight under the force of gravity; CFCs, for example, are large, heavier-than-air molecules which would otherwise remain close to the Earth's surface. (It is well that they do not since their global warming potentials are far larger than those of the natural greenhouse gases, see Part Three). As it is, CFCs are found everywhere in the atmosphere, where many of them will remain for longer than a human lifetimes (see Part Two).

The Meridional (north-south) Circulation is a slow circulation, taking many months to perform one cycle, but air also traverses the entire hemisphere, in the west-east direction in winter and the east-west direction in summer, in hardly more than a week. The chemical properties of trace gases such as ozone remain unchanged for months or even years, depending on altitude, making it possible to trace their paths in a chemical sense, but the dynamical properties of the atmosphere are transient, often prevailing for just a few hours. This restriction presented meteorologists with a serious problem: if they wished to trace the history of an air parcel, or predict its future whereabouts, how could they know they were tracing the same body of air throughout? Indeed, given how quickly gases mix when introduced separately into a container, was it really meaningful to speak of an air *parcel* at all?

The solution was quite ingenious: meteorologists contrived to find properties in the equations of motion and thermodynamics which did not change rapidly with time. Provided measurements of pressure, temperature and wind speed were available, such invariant quantities could be calculated. The two in most common use are called *potential vorticity* and *potential temperature*, customarily given the symbols **PV** and ϑ, respectively. A detailed description of them will be found in Appendix 1.6, but it is sufficient for the casual reader to know that they behave as if a parcel of air was tagged with a number, or coloured with a dye, which does not change appreciably over a period of several weeks, at least in the stratosphere. In the troposphere, where rapid vertical motions exist, these tags are only reliable for a few days, perhaps a week at best.

Knowing where air originated and predicting where it is going (the essence of weather forecasting) only gets us a little further forward, however. To really understand what is going on in the atmosphere, and to make predictions farther ahead than a few days or weeks, it is necessary to use tools which can assimilate vast quantities of information, and perform the millions of calculations required to achieve this. This task, once attempted by hand, now falls within the purview of computers and atmospheric models, the subject of the next chapter.

ATMOSPHERIC MODELS

During the 30,000 or so years since humans began to communicate their memories of the past via pictograms on cave walls, it has always been impossible, on the strength of observations of one's immediate surroundings alone, to predict with confidence how the atmosphere will behave even a day, still less a week, into the future. Throughout most of the twentieth century, despite having powerful theoretical tools at their disposal, atmospheric scientists have still been unable to calculate changes in weather patterns before they actually occurred. To some extent this is still true, of course, but things are changing. The advent of computers and the establishment of a geostationary and polar orbital satellite network have dramatically enhanced the accuracy of weather forecasts. Computers can assimilate the gigabytes of data obtained daily about the atmosphere and perform the millions of necessary calculations quickly enough to provide a meaningful prediction.

Given all this technology, one may ask why weather prediction is still not precise? One reason is that even the world's fastest computers, used by organisations such as the European Centre for Medium-Range Weather Forecasting (ECMWF) in Reading, are still not fast enough to solve all of the equations completely before the weather actually unfolds, so approximations (let's call them educated guesses) are still necessary. Faster computers will one day overcome this problem, but even then the outcome will still not be 100% accurate, because there exists a still more fundamental limitation to weather forecasting: the small-scale processes at work in the atmosphere.

The equations used for weather prediction in computer programs are those which work well for large-scales, whilst the smaller-scale processes are ignored, and for a very good reason: the coverage of measurements in the atmosphere is sporadic, even with a network of satellite-borne instruments orbiting the planet every day. Satellite instruments cannot reliably measure temperature all the way to the ground, and whilst radiosondes carried aloft by balloons are accurate, they are released primarily over land masses, and not all land masses at that. Little wonder that innumerable small-scale events are not incorporated into atmospheric models. Furthermore, we know little about what is happening in the lower atmosphere over the oceans which comprise 70% of the Earth's surface. These limitations mean that the computers must be programmed to approximate what we believe is happening using the equations governing atmospheric motion, which are themselves approximations. In other words, they have to guess. The cumulative uncertainties inherent in predicting future weather using limited data coverage still prevents our forecasts extending far beyond a week or so; longer than this and the predictions become highly dubious.

With these constraints in mind, try now to imagine the problems which arise when trying to predict what changes in atmospheric temperatures and circulation may do to the ozone layer in ten or twenty years time, or (and this is vastly more complicated still) what may happen to the climate fifty or a hundred years ahead. Still, we have to start somewhere, and as incomplete as our climate modelling skills may be, at least they are better than making ad-hoc predictions. The great strength of atmospheric and climate models is that, given time-averaged data which span decades or centuries, without those disruptive day-to-day fluctuations, they can successfully simulate the passage of numerous ice ages over the past few million years, for example. Extrapolating from past data, they also indicate that another ice age will occur in the not too distant future, about 5,000 years from now in fact, a prediction supported by a wealth of indirect evidence gleaned from various disciplines of science. Models may be used to emulate the fall and rise of the mean sea level as larger amounts of water are locked up as ice as an ice age begins, and released again when it ends.

The problem with climate predictions on the time-scale of a century, the duration of interest to scientists and policy makers when weighing the evidence for climate change, is that we are back in the realm of the short-term, small-scale changes. The controversial rise in the global-mean temperature in recent years, for example, may have a profound impact later on, but at the moment the magnitude of this rise lies close to the size of the uncertainties in our measurements.

The need to address climate change issues, discussed at length in Part Three, has led to huge advances in the development of computer climate models. But what, exactly, *is* a computer model? Clearly, it has nothing to do with the construction of the small-scale wooden or plastic ships, aeroplanes or spacecraft familiar to most of us as children. In science, a model is a sophisticated computer program, and in atmospheric science in particular, it is a program which makes use of all that is known about the circulation (the dynamics), the radiative balance (incoming and outgoing energy) and the chemistry of the atmosphere. Models which combine *all* of these properties and allow them to interact, instead of treating each of them separately, are known as *Coupled Models*. To gain some understanding of how such models work, we need to examine each component separately.

From chapter 1.1, you are aware that motion in the atmosphere arises from a number of competing forces which are, in an average sense, in a state of dynamical balance: gravity and pressure control its behaviour in the vertical direction; pressure and the Coriolis Effect in the horizontal. The set of mathematical equations which describe this behaviour, to a reasonable approximation, are used in a model to reproduce the observed atmospheric circulations (figures 3 and 4). Because they are nothing more than mathematical expressions, they can be converted into a language computer can comprehend, the necessary calculations being performed millions of times faster than they can be done by hand. For convenience, let's call this part of the computer model Module One; computer programs are often written in a *modular* fashion, enabling each part to be utilised separately, or in conjunction with the other parts.

Given a set of realistic initial conditions for the whole atmosphere (starting values for wind speed and direction, temperature, the position of the sun, and so on, defined for each point on the globe), this Module is capable of predicting the motion of the atmosphere for the near future. As a rule, this will not be especially successful for the more distant future because, as mentioned earlier, the input data are incomplete, necessitating the use of numerous approximations which generate cumulative errors (errors that become progressively bigger the longer the program runs). In addition to data limitations, there are still details about the physical processes we are trying to reproduce which remain poorly understood. Another constraint is the computer processing time (or CPU time) which is costly, and solving the full set of meteorological equations for a single snapshot of the atmosphere is CPU intensive; thus, various terms in the equations are neglected or approximated, depending on their relative importance.

The second major component in a climate model is the radiative (or energy) balance of the atmosphere, known in the scientific community as *radiative transfer* (Appendix 1.7), perhaps not the most edifying appellation. Basically, the driving force behind all weather and climate is the energy from the Sun, and averaged over a period of a year, we find that about a third of this incoming radiation is reflected straight back into space. Some of the remainder is absorbed by the atmosphere, but the bulk of it goes into the land and oceans at the surface of the Earth. The amount absorbed at the surface is nearly balanced at the top of the atmosphere by outgoing (re-radiated) radiation, now at infra-red wavelengths. The process of absorption by the ground and re-emission back to the atmosphere lengthens the wavelength of the light, which means energy is lost along the way. Whilst the incoming visible light is transparent to the gases in the atmosphere, the longer infra-red light radiated by the ground is not; it presents a larger cross-section to the atmosphere and some of it is absorbed. This process is shown schematically in figure 5a; for comparison, the full range of solar energy reaching the Earth (the electromagnetic spectrum) is presented in figure 5b, on which the regions most pertinent to ozone loss and climate change are indicated.

The balance of incoming and outgoing radiation can be expressed in a computer program by using mathematical equations called the Radiative Transfer Equations. These are really quite formidable in their complexity, and the treatment provided in Appendix 1.7 is only a simplified introduction; those interested in a more complete introduction to this subject should consult the bibliography for this part of the book for further reading. It is sufficient here to appreciate that it is possible, to a good approximation, to model these processes. By adjusting the energy balance slightly, which is essentially what global warming is all about (see Part Three), it is also possible to simulate the effects of future climate change, at least in theory. It will shortly become apparent, however, that matters are rather more complicated than this statement might suggest.

The third major component in a climate model is that which tries to simulate the chemistry of the atmosphere. Of the 92 naturally-occurring elements (the elements will be discussed in Part Two, in the context of the Mendeléev's Periodic Table), only a dozen or so are found in a gaseous form. Nevertheless, the number of possible reactions in the atmosphere is staggeringly large, making the modelling of these processes extremely difficult.

(A) Radiative Balance of the Atmosphere

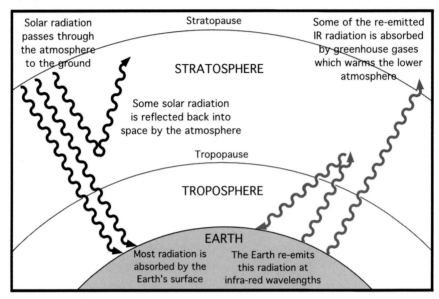

(B) The Electromagnetic Spectrum

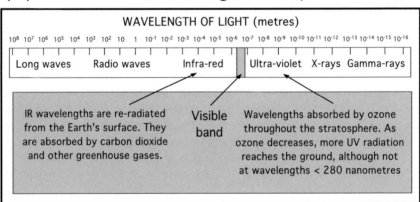

Figure 5 **A**, Schematic showing the natural balance of incoming and outgoing solar radiation in the Earth's atmosphere system; **B**, the electromagnetic spectrum showing the full range of wavelengths of the energy from the Sun. The ultra-violet radiation is important in the ozone issue; the infra-red in climate.

Many of these reactions are temperature dependent, and in consequence the rates at which they occur vary with altitude, season and latitude. Some only take place in daylight, others continue even at night. A model must incorporate both gas-phase (interactions between gas molecules) and heterogeneous (multi-phase) reactions, as well as all the microphysical processes involving small particles suspended in the atmosphere. It also needs to take account of condensation, evaporation, coagulation, sedimentation and the rate at which various chemical species are washed out of the atmosphere.

Modellers do manage to include hundreds, sometimes thousands of chemical reactions, both gaseous and heterogeneous types, but as numerous as these reactions are, they still only reproduce a small fraction of what goes on in the real atmosphere. Indeed, many models do not even account for the effects of the interaction between the ocean and the atmosphere, a key factor when considering climate change issues.

Putting the above modules together — the dynamics, radiative transfer and the chemistry of the atmosphere — the true complexity of making predictions about future climate becomes clear. To compensate for the limitations imposed by the models incompleteness, many chemical reactions and dynamical processes are parameterised in some way; if they are not, these omissions tend to make the model unstable; in other words, highly unrealistic results are likely to be generated.

An indication of the uncertainty in our predictive capabilities are the discrepancies which exist between different climate models. Fundamentally, they are all reproducing the same processes with only minor differences in their radiative and chemical schemes, but even so they often disagree quite dramatically. The problem arises because each of the models (and there are many) is parameterised in a different way, demonstrating their sensitivity to small changes when trying to make predictions about the climate as little as 50 years in the future.

Undoubtedly, models will continue to improve as our computers run still faster and our understanding of atmospheric processes continues to grow, but in the end, we may be faced with a limitation about which we shall be able to do nothing. All systems, from the behaviour of sub-atomic particles to the birth and evolution of the Universe, are inherently chaotic, and the atmosphere-ocean system on Earth is no exception. Ultimately, we shall be forced to live with a measure of uncertainty.

Appendices for Part One

APPENDIX 1.1

POWERS OF TEN

Because of the large distances traversed by air in the atmosphere, and the small numbers involved in discussing the number of atoms or molecules of a particular gas, it is advantageous to employ a mathematical shorthand to express them. For example, instead of writing out the present population of the human species, around six billion, as 6,000,000,000, it can be expressed as 6 x 10^9. This is called raising ten to a power, or using the exponent form. In other words, 100 (a ten followed by two zeros) is the same as 10^2, 1,000 the same as 1^3, and so on. Similarly, 26,900,000,000,000,000 can be expressed far more succinctly as 2.69×10^{16}. When moving to the domain of the very small, the mass of the hydrogen atom, 0.000 000 000 000 000 000 000 006 kg, becomes 6×10^{-24} kg.

The key here is to appreciate that the power to which ten is raised is *always* the same as the number of digits following the decimal point. For example, when 26,900,000,000,000,000 is expressed in exponent form, the decimal point is placed after the 2, after which follow 16 digits; similarly, for the mass of the hydrogen atom, 24 digits follow the decimal point.

When multiplying or dividing very large or small numbers, the rule is that for multiplication the indices are added, and for division they are subtracted. Thus, the speed of light (3×10^8 ms^{-1}) multiplied by a typical wavelength of visible light (3×10^{-7} m) would be as follows:

$$3 \times 10^8 \times 3 \times 10^{-7} = 3 \times 10^{(8+(-7))} = 3 \times 10^1 = 30$$

Similarly,

$$\frac{3 \times 10^8}{3 \times 10^{-7}} = 3 \times 10^{(8-(-7))} = 3 \times 10^{15}$$

For reference:

pico	10^{-12}	centi	10^{-2}	kilo	10^{3}
nano	10^{-9}	deci	10^{-1}	mega	10^{6}
micro	10^{-6}	deca	10^{1}	giga	10^{9}
milli	10^{-3}	hecto	10^{2}	tera	10^{12}

APPENDIX 1.2

THE RELATIONSHIP BETWEEN ATMOSPHERIC PRESSURE, TEMPERATURE AND GRAVITY

During the nineteenth century, a number of scientists sought to explain why rainy weather was associated with a drop in pressure at the ground, and why this decrease in pressure was usually linked to a corresponding drop in temperature. Experimenting with various gases, keeping pressure constant and changing volume or temperature, or keeping temperature constant and changing volume or pressure, finally led early investigators such as Boyle, Charles and Gay-Lussac to the following conclusion:

**Pressure multiplied by volume
and divided by temperature
will always give the same answer**

In mathematical terms, this can be written as

$$\frac{PV}{T} = constant$$

These symbols are the ones normally used to represent pressure (**P**), volume (**V**) and temperature (**T**). We call the constant **R** (the kinetic theory of gases later identified this as the molar gas constant, where a mole of gas is a quantity always containing the same number of atoms or molecules).

Volume is not a particularly useful quantity when investigating the atmosphere which, to all intents and purposes, is unbounded. Fortunately, one of the assumptions of the Ideal Gas Law above is that one can assume the inverse of the volume (that is, unity divided by volume, or 1/V) to be the same as the density of a gas, normally represented by the Greek letter ρ. After rearranging the Ideal Gas equation, it is possible to express pressure without recourse to volume:

$$P = \rho RT$$

This is the *meteorological form* of the Ideal Gas Equation. Both pressure and density can be measured, enabling us to deduce temperature (**R** is always constant). Similarly, if we measure temperature and pressure, density may be derived.

Vertically through the atmosphere, density and pressure are quantities that decrease exponentially. Exponential change (not to be confused with the exponent form discussed in Appendix 1.1) may be familiar from the decay of radioactive materials such as Uranium or Plutonium. For example, one kilogram of Plutonium has a half life (the time required for half of it to break down into a different element) of 26 years, so that after 26 years only 50% of it remains. After a further twenty-six years, only 25% (a quarter of a kilogram) will be left, and so on. Pressure and density in the atmosphere behave in a similar way: at a height of approximately 7 kilometres, 50% of the atmosphere remains above one's head; at 14 km, only 25%; at 21 km, 12.5%; at 28 km, 6.25%; at 35 km, 3.125%, etc. At an altitude of 100 km, just 0.006% of the atmosphere remains overhead, and we are now in interplanetary space.

This exponential behaviour enabled meteorologists to create yet another form of the Ideal Gas Law equation to yield information about altitude (**z**, in kilometres), rather than density. This version is known as the hydrostatic equation:

$$P = P_0 \times e^{-z/7}$$

where P_0 represents the pressure of the atmosphere at sea level (1000 hectopascals (hPa) in the Standard International, or SI, units). To use the **SI** unit for **z** in the above equation, the metre, substitute 7000 for 7. The quantity e, found on most calculators, is approximately 2.71828183.

By using this equation, we can calculate the pressure of the atmosphere at the summit of Mt. Everest, 8850 metres (8.85 km) above sea level:

$$P_{\text{AT THE SUMMIT OF EVEREST}} = 1000 \times e^{-8.85/7}$$
$$= 282 \text{ hPa}$$

APPENDIX 1.3

THE PRESSURE GRADIENT FORCE AND CORIOLIS EFFECT

The only important vertical forces in the atmosphere are the vertical pressure gradient force and the pull of gravity. The pressure gradient force exists because density decreases in the atmosphere with increasing altitude, so that the pressure at the surface of the Earth is a thousand times greater than at 50 km. This gradient is nothing more than a change in pressure over some vertical distance. In mathematical form, it may be expressed as $\Delta P/\Delta z$, where P is pressure, z altitude and Δ the Greek letter used to indicate change; for example, a drop in pressure from 1000 hPa to 980 hPa is 20 hPa; that is, $\Delta P = 20$ hPa.

Because pressure, like water flowing down a hill under the force of gravity, 'seeks' the lowest energy state possible, the atmosphere would flow out into space were in not for the force of gravity, g, which acts downwards towards the centre of the Earth. These two forces are in a state of balance. Mathematically, this is expressed as:

$$\frac{\Delta P}{\Delta z} = g$$

where $g \approx 9.81$ ms^{-2} at the surface of the Earth (the precise value depends upon latitude). Here, the power 2 indicates that this is an acceleration.

Similarly, the only important *horizontal* forces in the atmosphere are the pressure gradient force, which largely arises from differences in temperature (which, as explained in Appendix 1.2, is intimately connected with pressure) and the Coriolis Effect which is produced by the rotational motion of the Earth. For completeness, friction also plays a rôle, and this is briefly summarised in Appendix 1.5.

The pressure gradient force is defined as a variation in pressure over some distance. For example, when a cyclonic (low pressure) weather system is passing overhead, the atmospheric pressure just above the ground near its centre is lower than the pressure near the ground just outside of this region. Outside the cyclone, the pressure may be 1005 hPa whilst inside it is 980 hPa. This means that the pressure difference between the edge and the centre of the system is (1005 – 980) or 25 hPa. Since most cyclones are about 1000 km wide, the distance from the centre to the edge is around 500 km. The horizontal pressure gradient is therefore 25/500 or 0.05 hPa per kilometre (hPa km^{-1}). This is very small when compared to a vertical pressure gradient of ~150 hPa km^{-1} in the first kilometre above the ground, but it is still sufficient to produce strong winds.

This horizontal force, on large atmospheric scales, is balanced by the Coriolis Effect, the *apparent* deflecting force caused by the Earth's rotation which was first described in 1835 by the French civil engineer and physicist Gustave-Gaspard Coriolis (1792–1843). Coriolis showed that, if the ordinary Newtonian laws of motion of bodies (including, for example, air parcels, artillery shells and rockets) are to be used in a rotating frame of reference, then an inertial force acting at right-angles to the direction of the motion of the body must be included in those equations of motion. In the northern hemisphere, where the Earth's rotation is counter-clockwise, this apparent motion is to the right, whilst in the southern hemisphere (where the Earth's rotation is clockwise) it is to the left. In reality, the body does not really deviate from its path, but it appears to do so because of the motion of the co-ordinate system (the Earth) beneath it.

Looking at this in another way, consider a parcel of air some distance above the ground moving from near the equator to, say, 30°N. This air parcel is not rigidly connected to the surface of the rotating Earth: the atmosphere, remember, behaves in most important respects like a fluid. According to Newton's First Law, a body set in motion will continue in that motion until acted upon by some external, unbalanced force (although calling this Newton's First Law is not entirely correct since it was actually derived by Galileo Galilei). Neglecting friction and thermal effects, the air parcel will move in a straight line and with the same speed it possessed near the equator (about 463 ms^{-1}). At 30°N, the air parcel is still moving eastward at this speed, but the Earth's surface is now moving at only 403 ms^{-1}, because now it is closer to the axis of rotation. This difference of 60 ms^{-1} is the speed at which the air parcel is moving eastward relative to the ground, and an observer on the surface sees the air apparently deviating to the right of its original path.

The Coriolis Effect has a magnitude equal to twice the speed of rotation of the Earth multiplied by the speed of the parcel of air on which it acts. The Coriolis parameter, usually denoted by *f*, is written mathematically as

$$f = 2V\omega \sin (\lambda)$$

where *V* is the speed of the parcel of air, in metres per second (ms^{-1}), ω is the angular speed of the Earth's spin (in radians per second) and λ is the latitude of the parcel of air, in degrees. This requires further explanation. A radian measure (found on most scientific calculators) is used instead of degrees because of the sinusoidal function (of latitude) in the equation. In this form of measurement, the 360 degrees round a circle are the same as 2π radians (where $\pi = 3.141592$), so that 1 radian is the same as 57.2958 degrees. The number of degrees the Earth turns through each second is a small number; rotating 360° in twenty-four hours, it turns through

$$360 \text{ degrees } /(24 \text{ hours} \times 60 \text{ minutes} \times 60 \text{ seconds})$$

or 4.167×10^{-3} of a degree per second. This is the same as 7.2722×10^{-7} radians per second. For information about raising a number to a power, refer back to Appendix 1.1.

APPENDIX 1.4

THE ELECTROMAGNETIC SPECTRUM

The radiation from the Sun which bathes the Earth and makes life possible covers a very wide range of wavelengths, many of which cannot be detected by human senses. Most life on Earth has evolved to make use of the visible part of the electromagnetic spectrum; that part in which all the familiar colours (red, orange, yellow, green, blue, violet and innumerable intermediate shades) occur. We are aware, because of the sensation of touch, of the infrared part of the spectrum, commonly referred to as heat. We are also aware of the radiation in the ultraviolet range; a little more energetic than visible light, it is capable of burning our skin when we receive too much. Figure 5b demonstrates how small a part of the spectrum is accessible to our senses.

The speed of an electromagnetic wave (the speed of light, denoted c), irrespective of the part of the spectrum it occupies, is always constant (in a vacuum) at 299,792,458 metres per second (2.9979246×10^8 ms^{-1}). In air, it travels slightly more slowly. The *frequency* of the wave may be thought of as the number of wave crests (or oscillations) which pass a fixed point each second, and the distance from one wave crest to the next is called the *wavelength*. Frequency is normally denoted by the Greek letter υ, and is measured in a unit called *hertz*, which is defined as the number of oscillations of the wave per second. The relationship between the frequency of light, υ, and the wavelength of light, λ, (in metres) is

$$\upsilon = \frac{c}{\lambda}$$

Because of their applications during this century, we are aware of many other regions of the electromagnetic spectrum by their names, even if we cannot see them. Figure 5b shows the wavelengths from radio waves, infra-red, visible and ultra-violet radiation, right up to X-rays and gamma rays which, as well as coming from the Sun, are familiar through the radioactive decay of heavy elements such as Uranium and Plutonium.

The radiation arriving from the Sun covers a vast range, the longest wavelengths a million-billion-billion times larger than the shortest. The narrow band over which our senses perceive this radiation occupies only about $10^{-17}\%$ of the entire solar spectrum.

APPENDIX 1.5

THE EFFECTS OF FRICTION ON ATMOSPHERIC MOTION

The geostrophic wind, the balanced flow of air arising from the opposing tug of the horizontal pressure gradient force and the Coriolis Effect, is assumed to blow along isobars (lines of constant pressure), but this assumption neglects a very important force in the atmosphere: friction. Friction modifies the flow of air so that, rather than travelling along isobars, it tends to flow *across* them in the direction from high to low pressure. The angle between the isobar and the direction of the wind increases as friction increases, but as a rule of thumb, it is 30° ± 5°.

Near the Earth's surface, the wind flows to the left of the geostrophic wind direction, but as altitude increases and air density falls, friction correspondingly decreases, its influence already negligible at an altitude of about one kilometre. This first kilometre of atmosphere is known as the Boundary Layer, because it is strongly influenced by the frictional drag resulting from its interaction with the Earth's surface. It is a well-mixed region usually capped by a temperature inversion which prevents the air here easily mixing with the atmosphere above. Isolated in this way from the so-called *free* troposphere, pollutants emitted from the surface tend to remain near the ground, promoting the chemistry which gives rise to smog and the creation of high ozone concentrations.

APPENDIX 1.6

POTENTIAL TEMPERATURE AND POTENTIAL VORTICITY

Many of the trace gases in the atmosphere, which are important in ozone destruction and greenhouse warming, have lifetimes ranging from a few months to hundreds of years. Whilst it is possible to measure the amount of the shorter-lived gases at many altitudes, and use the concentration of the gas at each altitude to estimate how old it is, there is seemingly no way to determine from where the gas came in the atmosphere. For example, how might one trace the path of chlorofluorocarbons (discussed in Part Two) from the northern hemisphere mid-latitudes, where most of them are released, to their appearance over Antarctica? How can we label a body of air and follow its motion over days or even weeks?

This problem was solved by meteorologists by deriving two mathematical quantities relating the motion of the atmosphere which change only slowly with time. These quantities are Potential Temperature (usually denoted by the Greek letter ϑ) and Potential Vorticity (**PV**). Unfortunately, the meaning of neither of these properties is intuitively easy to grasp, and many of those who make use of them think about them in mathematical terms only. This is not surprising since they are both derived quantity, depending on a number of properties of atmospheric motion which are hard to visualise collectively. Because of their highly derived nature, both ϑ and **PV** suffer with the accumulation of errors in their component parts. Nevertheless, they work remarkably well over time-scales of a few days in the troposphere and a few weeks in the stratosphere, long enough to follow air over many thousands of kilometres.

Potential temperature is defined as *the temperature that a parcel of air would have if it were transported from some height in the atmosphere to a reference pressure (1000 hPa, approximately sea level) without exchanging any energy with its surroundings on the way.* To express this mathematically, we can start from the meteorological form of the Ideal Gas Equation in Appendix 1.2,

$$P = \rho RT$$

where **P** is the pressure, **ρ** the density, **T** the temperature and **R** the gas constant for dry air = 8.31441 (joules per mole per degree kelvin, or J mol^{-1}K^{-1}). Density can be eliminated from this equation by utilising the First Law of Thermodynamics for the case where no energy is exchanged between the gas and its surroundings (in other words, it behaves *adiabatically*). This law is expressed as follows:

$$c_p \frac{dT}{dt} = \frac{1}{\rho} \frac{dT}{dt}$$

where c_p is the specific heat of air at constant pressure, **dT/dt** the rate of change of temperature with time and **dP/dt** the rate of change of pressure with time. Solving for ρ,

$$\rho = \frac{c_p{}^{dP}/dt}{dT/dt}$$

Substituting this in the Ideal Gas Equation yields

$$\frac{P}{dP} = c_p R \frac{T}{dt}$$

Some of you may recognise this as a differential equation (differential calculus lies beyond the scope of this book) which can be integrated to obtain the mathematical expression for potential temperature:

$$\vartheta = T \left(\frac{1000}{P}\right)^{\frac{R}{c_p}}$$

Here, **P** and **T** are the pressure and temperature at some arbitrary height in the atmosphere, respectively.

But what is the advantage of using ϑ when trying to follow the path of an air parcel? By treating every air parcel as if it were at ground level, it effectively removes its vertical component of motion. This is akin to the altitude always being the same, so that, in effect, the air flows along on a surface of constant ϑ. Incidentally, potential temperature is measured in degrees kelvin (K), the absolute unit of temperature which starts at absolute zero (– 273.15°C) instead of the arbitrary freezing point of water. Thus, if air is at 400 K (corresponding approximately to an altitude of 15 km) today, it will still be at (or very close to) 400 K tomorrow. In the stratosphere where vertical motion is small in any case, this works very well indeed.

Having eliminated the vertical co-ordinate from the motion of the air, this is where potential vorticity (**PV**) comes into play. **PV** is even more highly-derived that ϑ and relies on no less than six properties of the atmosphere. It is the dynamical equivalent of a chemical tracer (a *dynamical tracer*, if you will), as indeed is ϑ. A chemical tracer is an atmospheric gas which remains unaltered for months (e.g., ozone in the lower stratosphere) of even for many years (e.g., nitrous oxide). Unfortunately, dynamical tracers do not remain accurate for such long periods of time, but like ϑ, **PV** is useful for up to several weeks in the stratosphere, (somewhat less in the troposphere).

Expressed mathematically,

$$PV = \frac{1}{\rho}\, \xi_A\, \frac{\partial \vartheta}{\partial P}$$

P is the density of the atmosphere, as before, and ξ_A is the *absolute vorticity*, a property intimately connected with the rotation of the Earth. The remaining part of the equation involves a combination of potential temperature and pressure which also requires a knowledge of differential calculus. Suffice it to say that $\partial \vartheta / \partial P$ is a measure of how stable the atmosphere is in the vertical direction, indicating how likely it is to mix with its surroundings.

For those interested in following the path of ozone through the atmosphere, there is a tremendous advantage to using **PV**, because it has an almost one-to-one relationship with ozone in the lower stratosphere. If ozone concentrations are low, **PV** is small; if ozone concentrations are high, **PV** is large. **PV** is often used as a surrogate for latitude because it identifies where ozone of a certain concentration originated; that is, it tells us whether the air in question came from high latitudes or from low latitudes. In each case, the amount of **PV** indicates how much ozone to expect, assuming no chemistry is involved. If, on the other hand, the amount of ozone is abnormally low for the calculated **PV** value, and we trace that air back to high latitudes in spring, it suggests the presence of ozone destruction by man-made chemicals.

Between them, potential temperature and potential vorticity take the place of vertical height and latitude, respectively, and are the keys to unravelling the history of the air.

APPENDIX 1.7

RADIATIVE TRANSFER

Radiative Transfer is a mathematical means of describing the exchange of energy between the earth and the atmosphere, or indeed the earth and the rest of the universe. The following description concentrates on the interaction between the absorption and re-emission of incoming solar radiation by the Earth's atmosphere, and how this relates to our planet's climate.

Quantum theory suggests that the energy transmitted (or transferred) by electromagnetic radiation occurs in discrete quanta, called photons. The rate of transfer is called the *radiant flux* which has units of joules per second, or watts (**W**). The radiant flux arriving from the Sun each second is 3.9×10^{26}W.

Dividing the radiant flux by the area through which is passes yields the *irradiance*, E (which is discussed in chapter 2.1 in connection with ozone). Because an area is now involved, the units are watts per metre-squared, or Wm^{-2}. If irradiance is considered for a particular wavelength (λ) rather than for the entire electromagnetic spectrum, it is called *monochromatic irradiance*, and because it is wavelength dependent, it acquires a subscript, E_λ. To obtain the irradiance over all wavelengths requires the use of integral calculus:

$$E = \int_0^\infty E\lambda \; d < \lambda$$

All bodies which radiate energy conform to a number of simple laws. The first of these is Kirchoff's Law, which states that the ratio of the irradiance for a given wavelength, and the absorbed radiation at the same wavelength, A_λ, depend only on the temperature of the body itself and the wavelength of the radiation. In other words, the nature of the body is irrelevant. This law can be expressed mathematically as

$$\frac{E_\lambda}{A_\lambda} = B_\lambda (T)$$

where $\mathbf{B}_\lambda$ $(\mathbf{T})$ is Planck's function. The bracketed $\mathbf{T}$ indicates its dependence on temperature. When $A_\lambda = 1$ for all wavelengths, the radiating body is said to behave as a *black body* which means that all incoming radiation is absorbed. When $A_\lambda < 1$, the radiating body is termed a grey body.

At a fixed temperature, $\mathbf{T}$, the energy radiated by a black body is a maximum for a certain wavelength, λ_{max}, given by Wein's Law (sometimes called the Wein Displacement Law):

$$\lambda_{max} = \frac{a}{T}$$

where a is a numerical constant $\approx 2.89 \times 10^{-3}$ m deg^{-1}. Wein's Law suggests that the Sun's radiation is concentrated in the visible and near-infra-red regions of the spectrum, whereas radiation emitted by cooler bodies such as planets occurs chiefly in the infra-red.

There is another important law which relates the intensity of the radiation emitted by the black body and its temperature: the Stefan-Boltzmann relation. In 1879, Stefan, using data supplied by an observational scientist named Tyndall, discovered that emissions at a temperature of 1473 K were 11.7 times greater than those at 798 K. Specifically, he noted that

$$\left(\frac{1473}{798} \right) \approx 11.7$$

which led to the formulation of the Stefan-Boltzmann Law:

$$E = \sigma T^4$$

where σ is the Stefan-Boltzmann constant $= 5.67 \times 10^{-8}$ W m^{-2} deg^{-4}. Historically, the formulation of Stefan-Boltzmann's law and Wein's law both precede the formulation of Planck's Law by some 50 years.

BIBLIOGRAPHY FOR PART ONE

The bibliography for books pertaining to meteorology is not extensive. The problem is that most text books are too advanced for the beginner, and most recent information is available primarily in scientific journals such as the Journal of Geophysical Research, Geophysical Research Letters, the Quarterly Journal of the Royal Meteorological Society and the Journal of Atmospheric Science. More widely available weekly journals which contain some useful information are Nature, Science and the New Scientist.

The books I have selected are designed for college students, but they are fairly accessible. They do require a solid background in physics and mathematics, however. If there is one disadvantage, it is that most these publications tend to be rather expensive.

BOOKS

Ahrens, C. Donald. "Essentials of Meteorology : An Invitation to the Atmosphere" (1997). ISBN: 0534537669. Approximate cost: £20.00

Goody, R. M. and Y. L. Yung. "Atmospheric Radiation: Theoretical Basis". Approximate cost: £35.00

Gordon, Adrian, Grace Warwick, Peter Schwedtfeger and Rola Byron-Scott. "Dynamic Meteorology: A Basic Course". Published 1998. ISBN: 0470244186. Approximate cost: £36.00

A Course in Elementary Meteorology, published by H. M. Stationary Office in conjunction with the U. K. Meteorological Office. *This is a very clear and thorough book for the beginner in meteorology.*

Lutgens, Frederick K., Edward J. Tarbuck and Dennis Tasa. "The Atmosphere: An Introduction to Meteorology". Published 1997. Approximate cost: £40.00 *This is a colourful tome and pleasant to read, perfect for the college student except for its price.*

Moran, J. M. and M. D. Morgan. "Essentials of Weather", 1994. ISBN: 0023838310. Approximate cost: £35.00.

O'Hare, G. "Atmospheric Systems: An Introduction to Meteorology". ISBN: 0050037420. Approximate cost: £18.00.
Published in 1940, this is an insight into the state of knowledge of the time.

PUBLICATIONS IN SCIENTIFIC JOURNALS

Kiehl, J. T. and S. Solomon. "On the radiative balance of the stratosphere." *Journal of Atmospheric Science, volume 43, pages 1,525-1,534*, 1986.

Trenberth, K. E. "The use and abuse of climate models." *Nature, volume 386, page 131*, 1997.

Part Two
The Ozone Layer

THE OZONE LAYER AND ULTRA-VIOLET RADIATION

The composition of the atmosphere has remained more or less constant for millions of years, providing a fairly stable environment in which living organisms have flourished. The primary gases are molecular nitrogen, making up 79% of the atmosphere, and molecular oxygen, making up 20%. The remaining 1% is comprised of a number of trace gases, all naturally occurring until recent times. Most trace gases do not occur as elements but as compounds, molecules formed by combining various elements. The elements are discussed in Appendix 2.1, and a modern version of Mendeléev's Periodic Table is provided in figure 6 for reference.

Familiar examples of compounds are carbon dioxide (one atom of carbon bonded to two atoms of oxygen) which is produced by respiration; carbon monoxide (one carbon atom bonded to one oxygen atom) released by incomplete combustion, and methane (one atom of carbon bonded to four atoms of hydrogen).

The abundance of trace gases has remained relatively constant for millions of years, although geological records reveal that the concentrations of some of them, especially the Greenhouse gases carbon dioxide and methane, have fluctuated on time scales of millennia. Since the advent of the industrial revolution, however, a number of trace gases (plus a suite of new ones which we have created) have increased in concentration, attaining levels higher than at any time in the past. The subject of increasing greenhouse gas concentrations will be developed further in Part Three. Here, we shall focus on just one trace gas (also a greenhouse gas, as it happens): ozone.

Since the mid-1970s, the normal abundance of atmospheric ozone has been severely affected by human activities. Like carbon dioxide and methane, ozone has increased in concentration in the troposphere, where an increase is undesirable because it is detrimental to human health. By far the large change in ozone

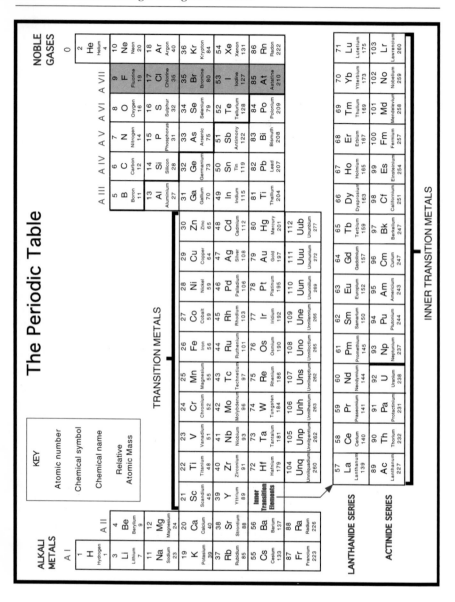

Figure 6 The Periodic Table of the Elements, showing the composition of all the naturally occurring elements, from hydrogen (the simplest and most abundant in the Universe) to Uranium (the heaviest and most unstable). The man-made elements, from atomic numbers 93 to 112, are included for completeness.

concentrations has occurred in the stratosphere, however, where it has declined at an alarming rate. Although only present in the stratosphere in concentrations of ~5 molecules for every million molecules of air, it nevertheless plays a vital rôle in maintaining life on Earth.

The ozone layer lies far above the summit of Mount Everest, which, you may recall, rises to an altitude of 8.85 km (or 29,000 feet) above sea level. Nearly all the ozone in the atmosphere is found in the stratosphere, the region between approximately 12 and 50 km where temperature remains fairly constant or increases with height. Ozone itself dramatically modifies the vertical temperature structure in this part of the atmosphere by absorbing short-wave (ultra-violet) energy from the Sun, and in so doing shields the Earth's surface from the effects of this biologically-harmful radiation. Without an ozone layer, it is unlikely that life on land would ever have developed.

Ozone, from the Greek word meaning *smell* (adopted because of its pungent odour) was discovered in 1839 by C. F. Schönbein as a by-product of electrical discharges, although its widespread presence in the atmosphere as a trace gas was not appreciated until 1850. Surface measurements of this gas began in the 1860s, which is how we know of its dramatic increase near the ground over the past century. The fact that an ozone layer existed in the stratosphere was not discovered until the early years of the twentieth century when experimenters found that it strongly absorbed light in the ultra-violet region of the electromagnetic spectrum (Appendix 1.4; figure 5B). Prominent amongst these early researchers was an Oxford scientist named Gordon Dobson who devised the first instrument to measure the total amount of ozone overhead, the so-called *ozone column*. The unit in which the size of this column is measured in called the Dobson Unit, in his honour (Appendix 2.2).

Until the last few decades of the twentieth century, the ozone layer was of interest only to a small number of scientists who studied this region of the atmosphere. As with most things which maintain our life-support system, the ozone layer was considered vast and indestructible, rather like the once-abundant equatorial rain forests which serve as the lungs of the Earth, and the seemingly endless oceans where we deposit so much of our waste. It came as a considerable surprise to many people that mankind really had the power to alter the environment on such a grand scale. We had grown accustomed to making changes *intentionally*, but the discovery that we were affecting the concentrations of a gas tens of kilometres up in the seemingly vast and boundless atmosphere was a big surprise.

¶The ozone layer has been thinning since the 1970s, although incontrovertible proof of this was not obtained until ten years later, in the mid-1980s. Moreover, the ozone layer is expected to continue thinning well into the next century. The chemicals responsible for causing ozone destruction, chlorofluorocarbons (CFCs) have been banned, but they are long-lived and we shall be forced to endure their legacy for many years to come. To worsen matters further, CFCs are not the only substances which contribute to ozone loss; a warming climate is also expected to enhance this process (chapter 2.4).

Let us look at what makes the ozone layer vulnerable to attack by certain chemicals. Ozone is actually a form of oxygen. The position of atomic oxygen in column A6 of the Periodic Table (figure 6) suggests that it has two free spaces in its outermost electron shell (Appendix 2.1), which means it cannot exist for long as a free atom. To attain the lowest energy configuration, achieved by the elements in column A8 of the Periodic Table by having full outer electron shells, atomic oxygen will readily combine with other elements or compounds to share a required pair of extra electrons. Sometimes, this other element is another atom of oxygen, in which case molecular oxygen forms. Occasionally, however, about a million times less frequently, that other substance is molecular oxygen and a molecule of ozone is created.

To describe succinctly the processes involved in natural ozone formation and loss, a shorthand has been adopted by chemists which uses the chemical symbols for each element shown in the Periodic Table. Atomic oxygen, the single atom of oxygen, is represented by 'O', while the oxygen we breathe (molecular oxygen) is written as O_2. (The subscript *2* denotes the number of atoms of a particular element which are present in a molecule). Similarly, ozone, comprised of three oxygen atoms, is written as O_3.

Using these chemical symbols and employing an arrow to indicate that a chemical reaction is taking place, we can express the chemical equation for the creation of ozone:

$$O + O_2 + M \Rightarrow O_3 + M \qquad [2.1.1]$$

This process, called *photolysis*, is shown schematically in panels 1 and 2 of figure 7.

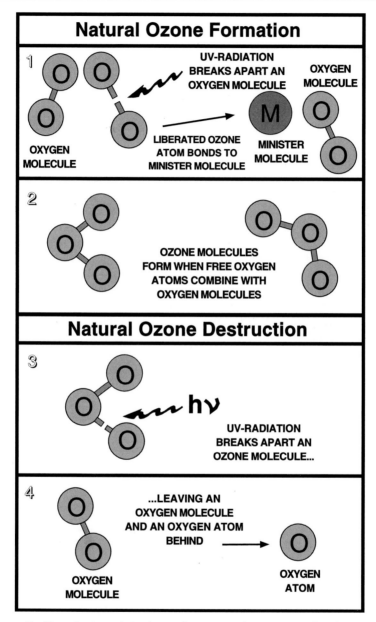

Figure 7 Panels 1 and 2 show the natural ozone production process in the atmosphere, whilst panels 3 and 4 show its natural destruction. These two cycles occur at similar rates, maintaining a constant amount of ozone in a time-average sense.

Notice how the number of oxygen atoms to the left of the arrow in equation 2.1.1 (three in total) is the same as the number to the right; all that happened as the reaction occurred was that oxygen was converted into ozone; no oxygen atoms were actually lost. All chemical reactions obey this rule; they are *never* destroyed. This is one of the fundamental axioms of science (these axioms being the laws of conservation of energy and matter). The *Law of Conservation of Mass*, for example, states that matter is conserved at all times.

In equation 2.1.1 is something called 'M', which appears on both sides of the arrow. 'M' is not an element which appears in the Periodic Table. It may be considered as any one of a number of different atoms or compounds which assist atomic and molecular oxygen to combine and make ozone by carrying away excess energy, but *without changing its identity*. In effect, these substances provide a viable energy pathway for certain reactions to occur, but this energy is only on short-term loan.

Not only is the ozone layer located in the stratosphere, it is also formed there. Solar radiation at wavelengths between 2.4–3.2 × 10^{-7} m (Appendix 1.1) is ultra-violet radiation (figure 5B). The process of ozone formation has two steps: (*a*) UV light is absorbed by an oxygen molecule (O_2) which breaks apart (dissociates) into two oxygen atoms, and (*b*) these oxygen atoms occasionally combine with another oxygen molecule to form ozone. As mentioned earlier, atoms combine together to achieve the lowest energy state possible, and once in this state, it may seem strange that a molecule like ozone should fall apart again. It does so because the very energy which enabled it to form, the UV light from the Sun, will also break it apart. By absorbing UV radiation in the wavelength range 2.4–3.2 × 10^{-7} m, the molecule acquires precisely the amount of energy required to overcome the chemical bonds linking its three component atoms, a process called *photodissociation*. Photodissociation may also be expressed in the form of a simple chemical equation:

$$O_3 + \mathbf{hv} \Rightarrow O_2 + O \qquad [2.1.2]$$

This process is also shown schematically in figure 7, in panels 3 and 4. The energy supplied by the radiation is represented by **hv**, where **h** is a constant and **v** is the frequency of the light (**hv** is described further in Appendix 2.3).

This set of chemical reactions was suggested in 1930 by a meteorologist named Sidney Chapman, just two years after Dobson identified the region between ~12 and 50 km as the seat of the Earth's ozone layer. To be entirely accurate, if the second reaction alone were responsible for the removal of ozone from the atmosphere, the ozone layer would be twice as thick as it is. There is, in fact, a whole family of chemical reactions involving the oxides of nitrogen (N) and hydrogen (H) (oxides are molecules made by combining other elements with oxygen). For example, combining nitrogen and oxygen can create nitric oxide (NO) or nitrogen dioxide (NO_2); similarly, hydrogen can combine with oxygen to make hydroxyl (OH) or hydroperoxyl (HO_2). It is these four gases (although not only these) which participate in the reaction to convert ozone back to oxygen, without themselves being affected.

In an un-perturbed state, ozone is formed and destroyed in the atmosphere in equal amounts (about one billion tonnes is naturally formed and destroyed every year), and this cycle is sustained because there is so much more molecular oxygen than ozone in the stratosphere, which in turn is far more abundant than the amount of atomic oxygen, providing a high probability of atoms and molecules colliding. These chemical reaction cycles are the reason why most UV radiation is screened from the Earth's surface.

The absorption of UV radiation by ozone deposits a considerable amount of energy in the stratosphere, keeping it significantly warmer than it would otherwise be. In effect, the ozone layer is responsible for the very existence of the stratosphere. Notice, however, that both the production of ozone (starting with an oxygen molecule being broken apart by UV radiation) and its destruction (involving UV radiation directly) are both dependent on sunlight, which means that both sets of reactions shut down at night. This has important ramifications in the polar regions in winter, where night lasts for around six months of the year.

Since ozone is created and destroyed naturally over the equator, it may seem surprising that it is present throughout the whole stratosphere from equator to pole, the overhead ozone column (Appendix 2.2) actually *increases* as latitude increases. The mechanism responsible for the observed distribution of ozone in the atmosphere is the presence of a Meridional Circulation which carries air from the tropical troposphere to the polar stratosphere, as described in Part One and shown schematically in figure 4. The ozone column is smallest in the tropics, despite its production there, measuring no more than 200–250 Dobson Units, or DU

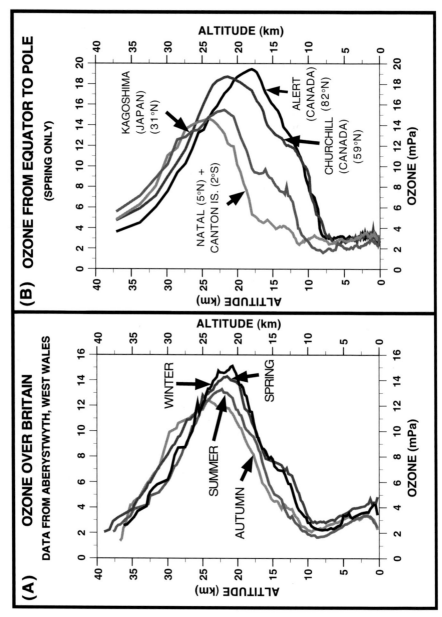

Figure 8 A, The seasonal variation in the vertical distribution of ozone above the U.K.; more ozone is naturally present during winter and spring than in summer and autumn. **B,** the variation in the vertical distribution of ozone between the tropics and the polar regions during spring.

(Appendix 2.2), and largest near the pole where it rises to 400-500 DU during the northern hemisphere winter and spring, and 350 DU during winter in the southern hemisphere. Of course, prior to the late 1970s, there was far more ozone in the total column over Antarctica during winter, typically 450 DU.

Ozone varies naturally with season (see figure 8A which demonstrates how it varies over the U.K.) and with latitude (figure 8B), the variation controlled by the strength of the Meridional Circulation. Because ozone molecules survive for months in the lower stratosphere (and for years in the middle and upper stratosphere), they are carried intact from their source region over the equator to higher latitudes. A lifetime of several months is sufficient for the journey to the polar regions in winter, promoting a large build-up of ozone there. As mentioned above, ozone production and destruction processes shut down in the absence of sunlight, allowing the incoming ozone time to accumulate. This is even apparent at mid-latitudes; compare the summer and winter profiles of ozone over Britain in figure 8A.

We shall explore the destruction of ozone by man-made means in the polar regions in chapter 2.2, and in mid-latitudes in chapter 2.3, but first, it will be instructive to examine the relationship between ozone and ultra-violet radiation in more detail. In particular, how does the amount of UV light reaching the Earth's surface varies with changes in the total ozone column?

UV radiation has been segregated into a number of sub-bands, called UV-A, UV-B and UV-C. UV-A radiation is that which is longer than 3.2×10^{-7} m, and is not absorbed by ozone at all; UV-B, with wavelengths in the range 2.8-3.2×10^{-7} m, lies on the lower edge of ozone's UV absorption band and therefore is intercepted by ozone (although not effectively), whilst UV-C, with wavelengths in the range 2.4-2.8×10^{-7} m is almost completely absorbed. A thinning ozone layer will therefore allow more UV-B radiation to reach the Earth's surface (see the schematic of this process in figure 9A). In theory, UV-C should also increase at the ground, but fortunately it is absorbed so efficiently in the stratosphere that a very large drop in ozone would be required for this to occur. If it ever does, the impact on the DNA and RNA of living organisms will be highly destructive.

One method to compare how much a given change in the total column of ozone above us will affect the amount of UV light reaching the Earth's surface is to examine a property of light called *irradiance*, the amount of radiation reaching the Earth's surface

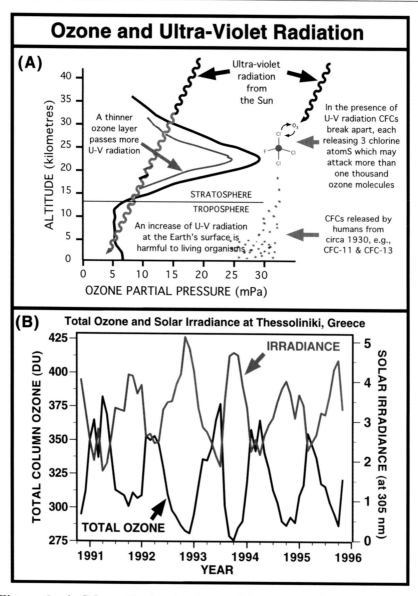

Figure 9 **A**, Schematic showing how a thinner ozone layer (caused, for example, by the release of CFCs) allows more UV radiation to reach the ground; **B**, variations in the ozone column and the amount of UV reaching the Earth's surface (irradiance) at 40°N. The relationship between ozone and UV irradiance is not one-to-one: a 1% decrease in the ozone column allows 1.5% more UV radiation to reach the Earth's surface.

at a given wavelength. Figure 9B shows the relation between total ozone and irradiance at 3.05×10^{-7} m (almost in the middle of the UV-B band) over northern Greece (~40°N) between autumn 1991 and autumn 1995. As the ozone column (black curve) rises and falls with season between approximately 270 and 380 DU, the irradiance (grey curve) mirrors the change almost exactly, dropping by up to 60% from the highest and lowest values. This effect varies with latitude since the UV-B radiation (and indeed all of the Sun's radiation) must traverse a progressively longer path through the atmosphere as latitude increases, and absorption becomes more efficient. The amount of UV-B radiation penetrating to the ground is also affected by clouds, but not in any straightforward and easily quantifiable fashion.

The large ozone losses over the south pole (chapter 2.2) mean that large increases in UV have been observed in Antarctica, despite the long path the radiation must take through the atmosphere at these high latitudes. The increase when the ozone hole is present is between 2 and 3 times higher than values measured prior to its existence, although UV-B amounts still remains largely unchanged at other times of the year when more ozone is present. There has also been an intermittent increase in UV-B at the extreme tip south of South America during periods when it is visited by the Antarctic vortex.

Elsewhere on the Earth, the increases in UV intensities at the ground are small, very much smaller in fact than the normal day-to-day variation. At the same time, such measurements are hampered by the brief period over which UV levels have been properly monitored, and also by an increase in ozone near the surface of the Earth arising from pollution. Observations over just a few years cannot be used to look for a long-term trend in UV irradiance, and we shall have to wait for at least another decade to see just how significant the changes in mid-latitudes will be.

A more detailed discussion on the detrimental effects of increasing UV radiation on living organisms will be found in chapter 2.4.

CHAPTER 2.2

OZONE LOSS IN THE POLAR REGIONS

It may seem, with the benefit of hindsight, that the world responded promptly and responsibly to avert the environmental crisis threatened by the thinning of the ozone layer. In truth, it started out as a war between science and industry, and arguments raged for a whole decade before something happened to settle the issue. The delay has proven to be costly, not to industry, but in terms of the length of time we shall now have to live with depleted ozone and its consequences.

In 1973, two American scientists, Professor F. Sherwood Rowland and his then research assistant Mario Molina, both working at the University of California, Irvine, decided to study the impact that certain man-made chemical compounds might have on the stratosphere. They focused on a family of chemicals which first came into existence in 1928: ChloroFluoroCarbons (CFCs).

CFCs were invented by a chemist named Thomas Midgley who worked for the American company General Motors. They asked him to create a safe refrigerant, and he came up with the first CFCs. These chemicals are a class of organic compounds which were subsequently used as refrigerants, aerosol propellants, foam blowing agents and solvents for the electronic industry. In the troposphere, they are chemically inert, a property which makes them ideally suited for the above purposes. Because of their many and varied applications, chemical companies such as Du Pont and ICI went into large-scale production of CFCs, and they quickly became a highly lucrative business. No one, least of all the chemical industry itself, wanted to have all this curtailed by scientific speculation about the potentially deleterious effects of CFCs when they reached the middle atmosphere.

Rowland and Molina theorised, in a scientific paper which they published in Nature in 1974, that CFCs could be broken apart by UV radiation, thereby liberating the element chlorine. Released from its electrochemical bond within the CFC molecule, the chlorine atom would then be free to react with other atoms and molecules in the atmosphere, including ozone. To affect ozone, however, they would need to reach altitudes of 30–40 km in the atmosphere. The net result, Molina and Rowland predicted, would be that the ozone molecule would be broken apart into its oxygen components, while the catalyst chlorine could go on to attack other ozone molecules, each time emerging unscathed from the chemical reaction.

It may seem far-fetched to think that large, heavy molecules like CFCs would ever get up as high as 35 km in the atmosphere, and this point was argued forcefully after the appearance of Rowland and Molina's paper. But those presenting this argument were ignorant about the Meridional Circulation discussed in chapter 1.2. This circulation is sufficiently strong to lift atoms and molecules *of all sizes*, irrespective of their mass, far up into the middle atmosphere; without it, molecules in the atmosphere would be gravitationally sorted by mass, the heaviest remaining close to the ground. As already indicated, however, this Meridional Circulation is not rapid, and compounds like CFCs which are released chiefly in the northern hemisphere mid-latitudes take weeks to arrive at the tropical tropopause. Once they enter the stratospheric arm of the circulation, their progress is an order of magnitude slower, taking up to five years to reach an altitude of 35 kilometres.

The second reason CFCs survive long enough to reach high altitudes is that they do not readily react with other molecules, making them difficult to destroy in the lowest part of the atmosphere where energetic radiation is less prevalent (see below). Their lifetimes are therefore long, some lasting for more than a century. Table 1 lists these atmospheric life-times, present abundance and rates of growth of a number of CFCs. In the troposphere, CFCs would have almost infinite lifetimes, and had they remained there, as the chemical industry once argued they must, they would now be making a very large contribution to global warming; CFCs are very effective greenhouse gases (chapter 3.4).

Molina and Rowland were later vindicated in their assertion that the reaction products of CFCs could destroy the ozone layer, but they were wrong about the precise nature of the mechanism which would give rise to this destruction. They reasoned in terms of normal, gas-phase reactions in the atmosphere (reactions which take place between the gases themselves without the assistance of a solid medium), which led them to anticipate a decrease in ozone of just a few percent over the next century. Nevertheless, even with these conservative estimates, there was still an outcry from the chemical industry as soon as their theory was published in the journal *Nature*. After all, if Molina and Rowland were taken seriously, the industry stood to lose financially.

Table 1   Halocarbon abundances and trends.

Halocarbon	Chemical symbol	Abundance (pptv) as of 1995	Annual rate of increase % as of 1995	Lifetime (years)
CCl_3F	(CFC 11)	280	4	65
CCl_2F_2	(CFC 12)	484	4	130
$CClF_3$	(CFC 13)	5	–	400
$C_2Cl_3F_3$	(CFC 113	60	10	90
$C_2Cl_2F_4$	(CFC 114)	15	–	200
$C2ClF_5$	(CFC 115)	5	–	400
CCl_4		146	1.5	50
$CHClF_2$	(HCFC 22)	122	7	15
CH_3Cl		600	–	1.5
CH_3CCl_3		158	4	7
$CBrClF_2$	(Halon 1211)	1.7	12	25
$CBrF_3$	(Halon 1301)	2.0	15	110
CH_3Br		10–15	15	1.5

As industry and science locked horns throughout the following decade, the anticipated ozone destruction was in fact already underway, its progress far more rapid than anyone had anticipated. The chemical reactions expected by Molina and Rowland were not the only ones to take place; rather than just interactions occurring in the gas phase, there were others which proceeded on *solid* particles, such as those provided by aerosols and ice. These were the so-called *heterogeneous* reactions.

Before discussing heterogeneous chemistry, let us look first at the evolution of the current theory of ozone destruction, and why it is CFCs break apart in the stratosphere when they have remained intact for so long.

Photodissociation of CFCs takes place only when the molecules reach an altitude where they are exposed to sufficiently short-wave (ultra-violet) radiation from the Sun. Too little UV radiation penetrates into the lowest part of the atmosphere to break them apart, ironically because the ozone layer absorbs it. Even in the middle and upper stratosphere (30–50 km above the ground), where the atmosphere is bathed in UV radiation, many CFC molecules will still not break apart at once. In fact, the probability of this happening is quite small, which is why they will prevail throughout the twenty-first century, even though production has been halted. The photodissociation of CFCs takes place in the very heart of the ozone layer, enabling the liberated chlorine atoms to react at once with ozone and convert it into oxygen. (This process is outlined schematically in figure 10).

At the end of a decade of bitter controversy, the opposition to Rowland and Molina's theory took a new turn when discovery of a *hole* in the stratospheric ozone layer above Antarctica was announced in 1985 by Dr. Joseph Farman and co-workers at the British Antarctic Survey. Analysing data from a ground-based instrument called a Dobson spectrophotometer located at the British Antarctic station Halley (76°S, 27°W), essentially the same instrument Dobson had designed over half a century earlier, one of Farman's students, Jonathan Shanklin, noticed that the total ozone column above the station was lower than usual during the early 1980s. When these results were finally published in Nature, Farman also found himself under attack by the chemical industry because he had proposed a causal connection between ozone loss and high levels of chlorine in the stratosphere. The fight did not persist for long, however; in 1986, the first American expedition (NOZE, the **N**ational **OZ**one **E**xperiment) set off for Antarctica to measure the extent of the ozone hole, followed the next year by a much larger campaign called the Antarctic Airborne Ozone Experiment (AAOE), charged with the mission of identifying of the chemistry which was causing the problem.

As an aside, it should be emphasised that the term *ozone hole* is rather misleading. Almost any episodes of low ozone, most of which are entirely natural, are generally referred to as an ozone hole, or mini-hole. See, for example, ozone minima, which are discussed in the next chapter. These mini-holes, if indeed one should even refer to them as such, are a hundred times less significant than the vast ozone hole which occurs annually in the south polar regions.

Farman and his colleagues were alerted to the existence of Antarctic ozone loss because they had data from their Dobson instrument dating back to 1957. Up until the mid 1970s, at least 300 DU of ozone were present in the overhead column (figure 11A), but gradually, this value began to fall. The spring-time decline was slow at first, but by the early 1980s it was becoming very noticeable, the column decreasing by almost 10 DU each year. By 1993,

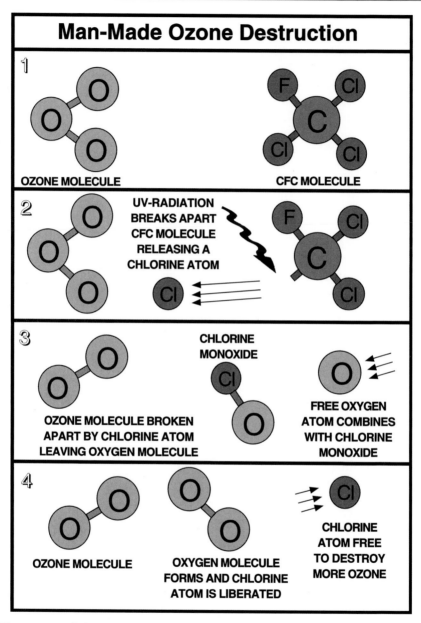

Figure 10   Schematic showing the man-made destruction of ozone. CFC molecules are broken apart by UV radiation and provide the necessary chlorine atoms for ozone destruction (via heterogeneous chemistry, not shown).

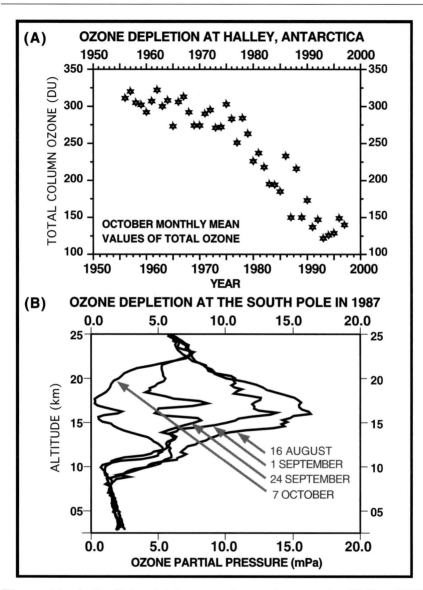

Figure 11 A, Declining total ozone column above station Halley (76°S, 27°W) in Antarctica from 1955 to 1998. The column has decreased by more than 50% during this period, and will continue to do so well into the 21st century; **B**, vertical ozone profiles over the south pole. The destruction process works very rapidly after the sun rises in spring when most of the ozone between 12 and 23 kilometres is destroyed in about one month.

the total October column was no more than 120 DU, just 40% of its original value. These figures are monthly averages; on individual days in recent years, the column has fallen below 100 DU.

This large, annual loss of ozone over Antarctica begins when the Sun rises as the end of winter, providing the energy to power the catalytic chemistry which so efficiently eats away at the ozone layer. In a matter of weeks, most of the ozone between 12 and 23 km is destroyed, as illustrated for the south polar station Amundsen-Scott in figure 11B.

The diminishing ozone column over the Antarctic should also have been seen by the satellite equivalent of the Dobson spectrophotometer, the Total Ozone Measuring Spectrometer, or TOMS. Such instruments have been orbiting the Earth since the late 1970s (there were even some earlier versions operating as far back as 1971). In fact, TOMS actually *was* recording the evolution of the ozone hole, but for years no one knew, and the researchers at NASA who were responsible for the orbiting TOMS instrument missed out on the atmospheric discovery of the century.

The reason NASA scientists missed seeing the hole develop was due to a short piece of computer code. This code, designed to impose quality control, led the program to *think* it was examining bad data; total column ozone values below 180 DU were considered to be erroneous since they had never been observed by ground-based or balloon-borne instruments. Thus, every spring as Nimbus 7 (the satellite platform bearing a TOMS instrument from 1979 until 1993) passed over Antarctica and recorded exceptionally low ozone values, the computer program threw them away. Fortunately, it was not discarding the data in a physical sense, otherwise there would be no satellite record showing the hole's development from space, as presented in the top six images in **plate 1**.

But mysteries remained. Why, for example, did most of the ozone loss occur between 13 and 23 km when, according to theory, most of the CFC chemistry was taking place between 30 and 40 km? Why too did it tend to occur chiefly in the polar regions, and then most strongly in the southern hemisphere? No one was sure of the answers when the ozone hole was first observed, demonstrating how little we knew about the atmosphere as recently as the mid 1980s. Nor should one infer that we know so very much more now; our knowledge is growing, but there are still innumerable processes we do not properly understand, and there are certainly still more we have yet to discover.

In Part One, it was shown how a jet stream forms in the polar

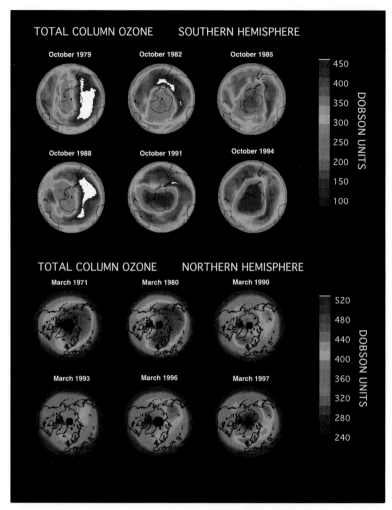

Plate 1 Polar ozone loss seen by orbiting satellites. Data from a type of instrument called a Total Ozone Measuring Spectrometer (TOMS), which has been mounted on various satellites since the 1970s, shows how the ozone column has declined not only over the south polar regions between October 1979 and October 1994, but also in the Arctic where the loss first became noticeable around 1990. Note that the colour scale for the top six images (southern hemisphere) runs from 100 to 450 DU, whereas for the bottom set of six images (northern hemisphere), it runs from 240 to 520 DU. There has always been more ozone in the Arctic spring than in the Antarctic because of the different dynamical behaviour of the atmosphere in each hemisphere. (*Courtesy of Mark Schoeberl and Paul Newman, NASA*).

stratosphere every winter (figure 3), its boundary forming a belt of high-speed winds in the vicinity of the polar circle, poleward of which the Sun does not rise during the winter. Recall that this circumpolar vortex effectively isolates the air trapped inside it from the rest of the atmosphere, allowing it to cool to very low temperatures, often below −80°C. Because the vortex extends far up into the atmosphere, it encloses a large volume of air, some of which was in the troposphere months earlier, and which therefore contains pollutants emitted by mankind, including the components of the CFCs by now photodissociated by ultra-violet light.

Because temperatures inside the Antarctic polar vortex often fall below −80°C, equivalent to approximately 193K on the thermodynamic scale, certain types of clouds are able to form. These clouds are composed largely of nitric acid and water, believed to be primarily in the form of crystals of nitric acid trihydrate, or NAT, chemical formula $HNO_3.3(H_2O)$. These are Type I PSCs. A second type (Type II) forms when temperatures reach −85°C (188K). Polar stratospheric clouds may also form from ordinary water ice, but these are by far the least common of the two. Such low temperatures are necessary for these clouds to condense because the stratosphere is extremely dry, containing less than 5 ppmv of water vapour, compared to several thousand parts per million in the troposphere.

PSCs are not a recent development; they have been observed for at least a century, variously known as nacreous or "mother-of-pearl" clouds, because of their vivid colours. It was not until 1982 that McCormick et alia (see bibliography) reported abnormally-high aerosol extinction values in the stratosphere observed by the Stratospheric Aerosol Measurement (SAM) II satellite sensor during the polar winters, and the true numbers and geographical extent of PSCs were mapped for the first time. This discovery was an important step in later unravelling the ozone depletion mystery.

Under normal conditions, most of the chlorine released by CFCs into the stratosphere is locked-up in harmless chemical compounds, the most important being chlorine nitrate (chemical formula $ClONO_2$) and hydrogen chloride (HCl), chlorine's so-called *reservoir* species. In the presence of PSCs, however, everything changes: polar stratospheric clouds provide a solid medium for heterogeneous chemical reactions to take place, allowing the chlorine reservoir compounds to react with other compounds and release chlorine atoms:

$$ClONO_2 + HCl \Rightarrow Cl_2 + HNO_3 \qquad [2.2.1]$$

Here, chlorine nitrate ($ClONO_2$) and hydrochloric acid (HCl) are converted into molecular chlorine (Cl_2) and nitric acid (HNO_3). Notice, as shown in the previous chapter, that the number of atoms of each element present on the left side of the reaction arrow is equal to the number on the right hand side; they have simply been broken apart and reassembled in a different order. Another pathway for releasing chlorine occurs if chlorine nitrate reacts with water (H_2O):

$$ClONO_2 + H_2O \Rightarrow HOCl + HNO_3 \qquad [2.2.2]$$

creating hypochlorous acid (HOCl), as well as nitric acid. (In reality, there are more than 100 chemical species involved, but an exhaustive treatment is far beyond the scope of this book; refer to Appendix 2.4 for more details, however).

Despite the complexity of these processes, the rate at which ozone is destroyed always depends on how much chlorine is present in the form of chlorine atoms (Cl), or in the form of chlorine monoxide (ClO). An important point to appreciate is that there is no more chlorine over Antarctica than anywhere else in the atmosphere; it is purely the unusual physical conditions in this region (the widespread presence of polar stratospheric clouds) which give rise to such efficient ozone destruction.

The molecules Cl_2 and HOCl are *reactive* rather than reservoir (*un-reactive*) forms of chlorine, which means they can be easily broken apart (*photolysed*) by gas phase chemistry. Even so, this can only happen if sufficient energy is present, which of course is supplied in abundance when the Sun rises over the Antarctic in spring, bathing the atmosphere in ultra-violet light. The returning Sun weakens the strength of the winds circling the pole, and ultimately causes the vortex to break down altogether, but the vortex manages to remain stable for several months after sunrise, keeping the air inside trapped and allowing ozone-destruction cycles to run amok, as follows:

$$Cl_2 + \mathbf{h\nu} \Rightarrow Cl + Cl \qquad [2.2.3]$$

Both chlorine atoms are now free to react with ozone, breaking it apart to leave oxygen (O_2) and chlorine monoxide (ClO):

$$Cl + O_3 \Rightarrow ClO + O_2 \qquad\qquad [2.2.4]$$

leading to

$$ClO + 2O_3 \Rightarrow ClO + 3O_2 \qquad\qquad [2.2.5]$$

which is the main polar reaction. At low polar stratospheric temperatures, these reactions are extremely fast and dominate the ozone depletion process. Moreover, a single chlorine atom can destroy tens of thousands of ozone molecules before it is permanently removed from the atmosphere.

Another member of the halogen family, bromine, is even more efficient than chlorine at destroying ozone because it has no stable reservoir species to sequester it away, which means it is *always* available in a reactive form. The saving grace with bromine is that it is about 100 times less abundant than chlorine, present in concentrations of only 30 parts per trillion (thirty bromine atoms for every 10^{12} atoms of air). This is just as well because, as scarce as it is, bromine still manages to act in concert with chlorine in a reaction to destroy ozone:

$$ClO + BrO \Rightarrow Br + Cl + O_2 \qquad\qquad [2.2.6]$$

which frees bromine and chlorine and enables them to attack ozone:

$$Br + O_3 \Rightarrow BrO + O_2 \qquad\qquad [2.2.7]$$

and

$$Cl + O_3 \Rightarrow ClO + O_2 \qquad\qquad [2.2.8]$$

This is the basis of the CFC theory, although these days it is rather more than just a theory. The scientists who went to the Antarctic in the mid-1980s, did much to confirm its premise, pushing aside explanations which suggested that ozone loss phenomena did not involve chemistry at all, surely music to the ears of the chemical industry, for despite the evocative graph published by Farman et alia showing a strong anti-correlation between ozone and CFC products, no one at that time had *proof positive* that these were the culprits. Conclusive evidence was finally forthcoming when an aircraft called the ER-2, a U-2 spy plane converted by

NASA for high altitude research, measured a whole suite of chemicals simultaneously inside the Antarctic polar vortex. Key amongst them were the concentrations of HCl and $ClONO_2$ (the reservoir species for chlorine), which were found extensively depleted. In exactly the same places where the reservoir species were missing, the concentrations of the active chlorine compounds ClO and HOCl were strongly enhanced. Moreover, ozone was depleted in almost precise anti-correlation with the variations in ClO, even on very small scales (a few hundreds of metres). The concentrations of ClO and ozone from one of these early flights are presented in figure 12.

As the ER-2 aircraft flew to high southern latitudes (going left along the horizontal axis on the graph) in September 1987, it recorded a cross-over in the concentration of ClO and O_3 as it crossed the edge of the circumpolar vortex; as the ozone levels plummeted, ClO increased by a factor of ten. Equatorward of the vortex, ozone levels were normal and ClO concentrations were small. This emphasised the crucial role played by this spinning vortex in promoting the chemistry of ozone destruction.

Since that time, numerous satellite instruments have been designed to measure not just these and also other important species in the polar stratosphere. **Plate 2**, from left to right, shows the strong relationships between temperature, nitric acid (HNO_3), chlorine monoxide and ozone: low temperatures are seen in the same region as elevated HNO_3 (a by-product of the ozone destruction process) as well as elevated ClO, which both correspond to the region of depleted ozone, here shown in parts per million by volume (parts per billion divided by a thousand).

The lost ozone cannot be replenished by air flowing horizontally in from lower latitudes because such movement in the atmosphere is halted by the boundary of the circumpolar vortex, leading to a build-up of ozone which forms a collar around its periphery. Actually, some air *is* exchanged across the vortex boundary, but the amount is believed to be insignificant compared with the billion or so cubic kilometres of air contained inside it. Most of this contained air does not escape until the circumpolar winds weaken in response to rising temperatures at the end of October over Antarctica (equivalent to the end of April in the northern hemisphere), at which time the vortex finally breaks down and allows the ozone-poor air from within to flow out over mid-latitudes, temporarily diluting the ozone shield over most of the hemisphere. The presence of the subtropical jet stream prevents most of this

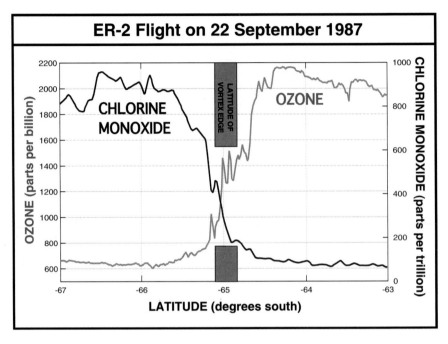

Figure 12 In-situ measurements of ozone (grey curve) and chlorine monoxide (black curve) measured by NASA's ER-2 aircraft as it flew south into the Antarctic polar vortex on 22nd September, 1987, obtaining the first firm evidence that humans was responsible for the ozone hole. Note how O_3 concentrations fall and ClO concentrations rise sharply as the aircraft crossing the vortex boundary at 65°S (going from right to left on the figure).

air from reaching the equatorial regions where almost no ozone loss has been observed.

The chemistry which eats away at the ozone layer inside the polar vortex has been successfully reproduced under laboratory conditions using a pressure chamber, adding to the evidence that chlorine reservoir compounds react on the surfaces of polar stratospheric clouds (PSCs) to release chlorine. Moreover, airborne measurements demonstrate that air which has passed through regions occupied by PSCs is low in chlorine reservoir species, containing instead large abundances of the active chlorine compounds.

Earth's Lower Stratosphere in 1996 Northern and Southern Winters

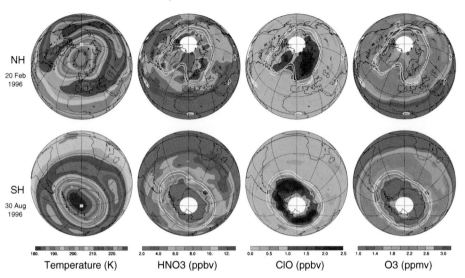

NH
20 Feb 1996

SH
30 Aug 1996

Temperature (K) HNO3 (ppbv) ClO (ppbv) O3 (ppmv)

Plate 2   Satellites are now able to measure numerous chemical species in the atmosphere, and provide additional evidence for the human impact on the ozone layer. As well as temperature, the column amounts of nitric acid (HNO₃, an important intermediary in the ozone destruction process), chlorine monoxide (ClO) and ozone (O₃) are recorded simultaneously in both the northern (top four images) and southern (bottom four images) hemispheres. Notice how areas of low temperature, HNO₃ and O₃ correspond closely to the area occupied by high concentrations of ClO. (*Courtesy of Joe Waters, NASA JPL*).

In the years following its discovery, the Antarctic ozone hole has grown steadily bigger, reaching its maximum areal extent to date in October 1998, now filling the entire Antarctic vortex. The vortex boundary is at times distorted by atmospheric disturbances, elongating it towards the equator so that it passes over the tip of South America, and parts of Australia and New Zealand. In general, however, the vortex remains confined to latitudes greater than 55°S, and should remain so unless the Antarctic wind patterns change substantially. In Part One, it was explained how temperature and wind speed are connected; in Part Three, this will be considered again in the context of a shift in climate which may alter wind strengths and patterns over much of the globe.

Once the mechanisms responsible for the Antarctic ozone hole had been understood, questions inevitably arose as to the possibility of similar losses at northern high latitudes. This was quickly quashed by the realisation that the flow patterns of the winds in the Arctic differ substantially from those in the Antarctic. The rotation of the northern hemisphere polar vortex is perturbed far more frequently than its Austral counterpart, warming the stratosphere and, for the most part, preventing the very low temperatures which trigger the formation of PSCs. Without PSCs, chlorine is not released from the inactive (reservoir) compounds and ozone destruction will not take place.

What is it that disrupts the Arctic polar vortex? In Part One, there was mention of how mountain ranges disturb the flow of the atmosphere, and the northern hemisphere boasts numerous mountain ranges, in contrast to the southern hemisphere where there are few. The waves set up by the Earth's topology propagate far up into the atmosphere, depositing large amounts of energy in the stratosphere. At times, this deposition of energy is sufficient to slow down, or even reverse the direction of flow of the vortex winds. When this happens, as it does in most winters, the entire vortex to likely to break apart. If such an event, known as a sudden stratospheric warming, takes place in early or mid-winter, the vortex will usually re-establish itself, but if spring is near, the disruption is often permanent and the vortex is dissipated early, allowing far less time for heterogeneous chemistry to occur even if PSCs had been present.

The reasoning seemed sound enough. But then, in the early 1990s, ozone depletion began occurring in the Arctic on a grand scale, still not as dramatic as inside the Antarctic polar vortex, but the loss was still significant. **Plates 1** and **2** both show that Arctic ozone loss is taking place. But why? In the 1980s, there were only a few days during each winter when Arctic temperatures fell far enough for PSCs to form, and even then their distribution was very patchy, but during the 1990s, Arctic temperatures have generally been lower, setting up a more robust polar vortex which persists for longer. The stratosphere, it seemed, was cooling, perhaps in response to some natural phenomenon (as yet undiscovered), or perhaps in response to an enhanced greenhouse effect (chapter 3.4).

It is now widely believed that as global warming heats up the troposphere, it will, at the same time, tend to cool the stratosphere (where the presence of more greenhouse gases will lead to more heat being radiated back to space). A colder stratosphere will in turn promote a more stable vortex, fostering the formation of more PSCs and hence more efficient ozone destruction. Ironically, as ozone decreases in the stratosphere, less ultra-violet radiation will be absorbed and temperatures will fall still further, strengthening the winds and making the polar vortex yet more stable, thereby enhancing heterogeneous chemistry which destroys more ozone, and so on. Global warming, then, may be setting up *a positive feedback mechanism.*

There is considerable uncertainty as to how severe ozone depletion in the northern hemisphere will become. The sequence of six satellite images at the bottom of **plate 1**, showing how ozone concentrations have declined during the month of March between 1971 and 1997, suggests that the problem is indeed serious. **Plate 2** permits a direct comparison of the late winter losses in the northern hemisphere (in February, 1996) with those in the southern hemisphere (in August, 1996), although note that the total ozone column scales are not the same. The Arctic decrease in ozone is less dramatic than that seen in the Antarctic, but it is undeniably happening, and growing steadily worse. In March 1997, ozone in the Arctic reached the lowest levels ever recorded, around 240 DU. For the first time, concentrations previously found only at sub-tropical latitudes (where they are naturally low) were observed close to the north pole (the purple colour in **plate 1**), in stark contrast to the high (red) concentrations (460–480 DU) present during the 1970s.

One particular question has been raised a number of times since ozone depletion was discovered: why should we be concerned about low ozone in the polar regions when hardly anyone lives there? Quite apart from the numerous life forms which inhabit the land and surface waters at high latitudes, for whose well-being we must surely accept responsibility, the problem is actually not confined to high latitudes. As the meteorology described in Part One makes abundantly clear, air which is today over the Arctic may well be over mid-latitudes in less than a week. Moreover, the north polar vortex is frequently displaced from the Arctic by atmospheric wave activity, pushing it equatorward where is passes over Russia, Europe and Canada. More significantly, when the polar vortex breaks down in spring, the air it contained is free to move over populated areas and thin the ozone layer (*the dilution effect*), causing a dramatic if transitory decrease in mid-latitude ozone columns. If the dilution effect ever became so severe that the mid-latitude ozone layer did not fully recover before the following winter, then its effects would become cumulative. Whilst the currently-held opinion is against such a protracted effect, it would go some way towards explaining the ozone losses which have been recorded over mid-latitudes in recent years, currently thought to be 3–6% per decade.

OZONE LOSS OVER
MID-LATITUDES

As winter draws to a close and the Sun rises over the polar regions, the vortex breaks down completely. As mentioned in the previous chapter, this allows an outpouring of ozone-poor air into mid-latitudes which serves to dilute ozone concentrations over much of the hemisphere, so that temporarily the ozone layer is thinned by 5–10% over the most heavily populated regions on Earth. This may not seem serious; after all, in the months that follow, air from the tropics, richer in ozone, will move in to replenish it, and provided the time taken to replenish mid-latitude ozone levels does not last until the following winter, there may seem little cause for concern.

But then, another surprising discovery was made. Data from both ozonesondes and satellite-borne instruments revealed that, over the past fifteen or so years, the ozone layer in mid-latitudes has been getting steadily thinner every year, *throughout* the year. As in the polar regions, the greatest losses were still occurring in late winter and spring, but it was also happening in the other seasons. Table 2 indicates how the amount of depletion varies as a function of both season and latitude.

As in the early days of polar ozone loss, we know that mid-latitude ozone loss is taking place, but we don't really understand why. As usual there's no shortage of theories, but so far none of them offers a satisfactory explanation to account for *all* the observations.

One of the difficulties we face in trying to quantify mid-latitude ozone loss arises because it is so much smaller than polar losses, where a decrease of 50% is observed in matter of weeks. The situation is further complicated by the large, natural seasonal variation in ozone (as shown in figure 8) which tends to swamp the signal of man-made loss. To isolate a possible trend, it is first of all necessary to remove every known, natural variation. Two example are: (*i*) changes in the total column of ozone in response

Table 2 Sub-polar, mid-latitude and equatorial stratospheric ozone trends.

Latitude	% January	% April	% July	% October
65°N	−3.0	−6.6	−3.8	−5.6
55°N	−4.6	−6.7	−3.1	−4.4
45°N	−7.0	−6.8	−2.4	−3.1
35°N	−7.3	−4.7	−1.9	−1.6
25°N	−4.2	−2.9	−1.0	−0.8
05°N	−0.1	+1.0	−0.1	+1.3
05°S	+0.2	+1.0	−0.2	+1.3
25°S	−2.1	−1.6	−1.6	−1.1
35°S	−3.6	−3.2	−4.5	−2.6
45°S	−4.8	−4.2	−7.7	−4.4
55°S	−6.1	−5.6	−9.8	−9.7
65°S	−6.0	−8.6	−13.1	−19.5

to the 11-year solar sunspot cycle, during which the energy output from the Sun varies and alters the amount of ozone production and loss in the atmosphere (by several percent), and (*ii*) the change in transport caused by cyclic variations in the tropical winds (e.g., the quasi-biennial oscillation, or QBO) which affect the transport of ozone from its source region. Unfortunately, data from the various satellite instruments observing ozone loss do not agree on the magnitude of the decrease. They do agree on one thing however; they all detect a *decrease* in ozone everywhere except in the tropics. So far, not one has reported an increase.

Because of further disparities between satellite and balloon-borne instruments, a lot of time has been expended on tracking down the sources of error which have given rise to these differences, and although this issue is far from resolved, significant advances have been made. Figure 13 shows how ozone in mid-latitudes has changed in the lower stratosphere during the past few decades. The thick solid curve shows the size of the downward trend in ozone gleaned from mid-latitude ozonesonde data (acquired over the mid United States and Germany), whilst the thinner curve derives from ozone profile data collected by the SAGE (*Stratospheric Aerosol and Gas Experiment*) sensor mounted on NASA's Earth Radiation Budget Satellite (ERBS). The two types of instrument now agree to within 1%, and both show that ozone is decreasing throughout the lower stratosphere, most severely between 14.5 and 15.5 kilometres where the decadal

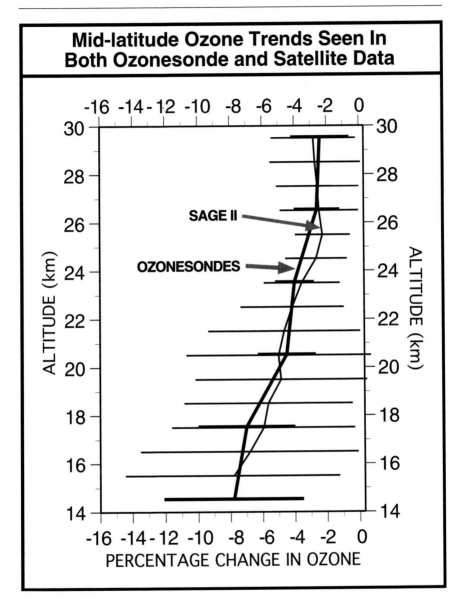

Figure 13 Ozone depletion is not confined to the polar regions. It has also been measured over mid-latitudes, both by satellite (SAGE II; thin curve) and balloon-borne instruments (ozonesondes; thick curve). The horizontal bars on these curves indicate the amount of uncertainty in the measurements.

decrease is almost 8%. You can read more about mid-latitude trends in ozone, as observed by satellite, in publications in scientific journals referenced in the bibliography for Part Two of this book.

A further difficulty in trying to detect human-induced ozone loss is to separate out chemical changes from purely dynamical effects. Atmospheric motion does not occur uniformly at all heights, so that vertically-deep slabs of the atmosphere, anywhere from a few tens of metres to more than 5 kilometres deep, can move horizontally over vast distances on time-scales of a few days.

The competing influences of chemical and dynamical effects may be compared using figure 14. Panel A shows a pair of vertical ozone profiles recorded at NyÅlesund (79°N), (a town on the island of Spitzbergen located deep inside the Arctic circle), one on 25th January and the other on 27th January 1992. On the 25th, the ozone profile in the lower stratosphere exhibited its characteristic shape for this latitude, increasing up to around 15–16 km and then dropping off with increasing height. In the profile two days later, however, most of the ozone between 12 and 20 km has disappeared. When this missing slab of ozone is compared to the pair of profiles in Panel B, recorded at McMurdo Station (78°S) in Antarctica on 19th August and 5th October 1994, the change looks remarkably similar. At first glance, then, it would appear that just as much ozone is being *lost* in the Arctic as in the Antarctic, and by the same mechanism.

But there are *three* important differences: Firstly, the sharp drop in ozone above NyÅlesund happened in just *two* days, whereas at McMurdo the profiles were obtained more than seven weeks apart. Secondly, the island of Spitzbergen is still locked in the darkness of the polar winter at the end of January, so there can be no solar energy to power heterogeneous chemistry. In contrast, the Sun has just risen at the end of August in Antarctica, and heterogeneous chemistry is quite certainly possible. Finally, the recent history of the air parcels containing low ozone is not the same in both cases.

Using a model of the atmosphere, it is possible to calculate the direction and speed of winds over the entire planet. These calculations are imperfect because, as discussed in chapter 1.3, the information fed into such a model is incomplete, promoting certain errors. However, on time-scales of a week or two, one can reliably trace the path of air backwards in time. Moreover, the derived invariant properties of air, potential temperature (θ) and potential vorticity (**PV**), (Appendix 1.6) enable a crude check on whether or

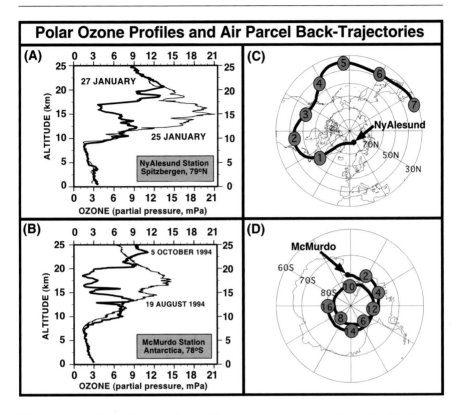

Polar Ozone Profiles and Air Parcel Back-Trajectories

Figure 14 Not all episodes of low ozone concentrations in the lower stratosphere are caused by heterogeneous chemistry. The low ozone events in (A) the Arctic and (B) the Antarctic look similar, but only the latter results from chemical destruction. Panels (C) and (D) show the paths the air has followed to each of these locations, respectively, demonstrating how the low ozone in the Arctic came from the subtropics (where ozone concentrations are naturally low) during the preceding week, whereas in the Antarctic the air never left the southern polar regions.

not the air parcel under investigation retains its integrity during advection (the name given to horizontal transport in the atmosphere).

To trace the air backwards in time, the trajectory calculations are initiated from the time, altitude, latitude and longitude of the observed low ozone events shown in panels A and B of figure 14. In Panel C, the history of the air which arrived above NyÅlesund on the 27th January 1992 reveals that, five days earlier, it was in the subtropics, and in the days before that was flowing more or less parallel the subtropical jet stream. Since there is no heterogeneous chemistry at work in the tropics or subtropics (although recent opinion suggests that there may in fact be some), we can be fairly confident that the air has not been chemically processed, and that it simply flowed poleward from a region where ozone concentrations are naturally low. The value of potential vorticity associated with this air parcel remained fairly constant with time, suggesting that it did not mix significantly with its surroundings en-route.

The back trajectory from McMurdo (panel D) tells quite a different story. Here, the air parcel simply encircled the pole, never travelling equatorward of 55°S. Throughout its recent history, it had remained inside the Antarctic polar vortex, and because ozone concentrations over the pole during winter and spring should be higher (known from data which pre-date the existence of the ozone hole), heterogeneous chemistry is almost certainly responsible for the missing ozone in the lower stratosphere.

Returning to the subject of mid-latitude ozone loss; low ozone episodes like the one seen in figure 14A are not confined to the Arctic, but also occur throughout the mid-latitude stratosphere. Figure 15A shows such an event over Britain on 7th March 1997. Might not such intrusions of ozone-poor air (let's call them *ozone minima*) from the sub-tropics contribute to the observed ozone loss trend in mid-latitudes? If this process is entirely dynamical, and has been in progress for centuries or perhaps millennia, its net contribution over a period of time should be zero. But if the number of ozone minima entering mid-latitudes has increased in recent years, then they may indeed contribute to the observed trend. Figure 15B shows the annual percentage of mid-latitude ozonesondes which have observed ozone minima (between January and April, inclusive) in the lower stratosphere since 1968 over Europe (black curve) and since the early 1970s over Canada (grey curve). It would appear that the percentage of minima passing over Europe has risen fairly dramatically during the last decade, rising from around 20% to more than 30%, although with considerable variability. The change over Canada is far less dramatic, although even these data do show an increase in some years, in concert with those seen in Europe·

The increase in ozone minima detected by ozonesondes is also reflected to some degree in the total ozone data retrieved from

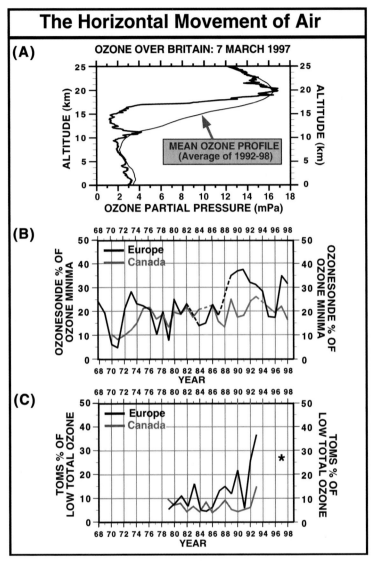

Figure 15 Low ozone episodes produced by the motion of the atmosphere occur in mid-latitudes too. Panel (A) shows such a minimum in the ozone layer over Wales on 7th March 1997 which, a week earlier, was in the subtropics; (B) the number of low ozone episodes as a function of time, showing how they have increased in abundance over the past thirty years, as shown by ozonesonde data from 1968 onwards; (C) occurrence of areas of low total column ozone seen by satellite-borne TOMS instruments from 1980 onwards.

orbiting TOMS instruments since 1979 (figure 15C). A total ozone column below 300 DU near the co-ordinates of the ozonesonde stations, a value somewhat below the seasonal average, was used to obtain the result. It is assumed that a low total column of ozone reflects the presence of an ozone minimum in the lower stratosphere, which is quite often the case, but not always since the height of the tropopause also changes in response to the presence of low of high pressure regions in the troposphere, which also affects the total column. Nevertheless, there is still some qualitative agreement between the curves in figures 15B and 15C, despite the fact that different quantities are actually being compared. The reason for this recent increase in the number of ozone minima and low total ozone columns over Europe is presently unclear, but it certainly has a dynamical component.

What does this mean for the mid-latitude trend in ozone? Research has shown that a typical ozone minimum covers an area comparable to, or bigger than, the size of Britain. The amount of ozone missing in the total column over this area is about 15% (compared to a climatological average), sufficient to noticeably affect the total amount of ozone present over the entire mid-latitude hemisphere. Moreover, ozone minima tend to persist for about a week, at times even longer, so that their presence is detectable even when measurements are averaged over an entire month. In the northern hemisphere between 40° and 60°N, minima may reduce the monthly-mean amount of ozone by up to 1%, creating a small *dynamical* trend which lurks beneath the larger, chemically-induced one 6–8%. It is noteworthy that the trend in ozone is largest at around 14.5 km (figure 13), the same altitude at which ozone minima are most prevalent.

The mystery of the year-round, mid-latitude ozone loss, when solved, will almost certainly turn out to have not a single source, but a number of smaller contributions which, taken in concert, add up to the observed trend. There are several more dynamical contenders, for example. Propagation anomalies have been detected in the monthly-mean atmospheric circulation in the tropics, disturbances which some researchers believe take between 5 and 10 years to reach middle and high latitudes. These anomalies began to manifest themselves in the early 1980s when, researchers at the National Center for Atmospheric Research in Boulder, Colorado, have suggested, a decadal shift in the amount of wave activity in the troposphere occurred.

Still another contributor to mid-latitude ozone loss may be the catalytic destruction cycle involving chlorine and/or bromine, very much as they occur in the polar regions. There has been much speculation about a tropical source for this type of chemistry, but so far there is no unequivocal evidence in support of it. At the same time, there was (and still is) a natural lag of 10–20 years between CFC emissions and their contribution to polar ozone depletion, so that events in progress today may not show up for several more decades to come.

THE FUTURE OF THE OZONE LAYER

There is widespread opinion now that the worst of the ozone destruction crisis is over, and that the ozone layer is well on the way to a complete recovery. But that is not entirely true; quite apart from the fact that the ozone layer will not return to 1980 concentrations for at least half a century, another force is at work in the atmosphere which may delay that recovery by several more decades. Rather than being over, the worst may be yet to come.

In the following pages, we shall examine some of the potential dangers which still lay ahead, and follow the evolution of the Montreal Protocol, the legal document drawn up during the 1980s (and which is still evolving as new scientific discoveries are made) in an attempt to protect the Earth's ozone layer. It has certainly been successful in bringing about substantial changes, but ironically the remedies it proposes to safeguard stratospheric ozone are expected to have a deleterious effect on the enhanced greenhouse effect.

(i) The Impact of Increasing UV Radiation at the Earth's Surface

In chapter 2.1, the relationship between ozone and ultra-violet radiation indicated how UV light both creates and destroys ozone via natural processes which have been operating in the atmosphere for many millions of years, and how absorption by the ozone layer protects the Earth's surface from excessive exposure to UV. Probably the mostly widely publicised aspects of the UV problem is its relationship to skin cancer.

Most skin cancers fall into three categories: basal cell carcinomas, squamous cell carcinomas and melanomas. Around 0.2% of the population of Britain and nearly 0.3% of the population of North America (100,000 and 750,000 people, respectively) contract one of these three forms of cancer every year. Fair-skinned people of north European extraction are particularly susceptible. In the case of North America, more than 90% of skin carcinomas are attributed to UV exposure, although this value (like UV radiation itself) varies with latitude, the incidence of cases declining from equator to pole.

The way in which UV induces the carcinoma (non-melanoma) form has been identified: in our DNA, the pyrimidine bases form dimers (duplicates of themselves, but with the information not necessarily stored in the correct order) when they absorb UV-B. This causes transcription errors as the DNA replicates itself, giving rise to genetic mutations. These types of cancer are easily treated if they are caught early enough.

Scientists at the National Center for Atmospheric Research in Boulder, Colorado, (Madronich and de Gruijl) have modelled the expected increases in non-melanoma skin cancer due to ozone depletion over the period 1979–1992: these are summarised in Table 3.

Table 3 Expected increases in skin Carcinoma rates due to ozone depletion over the period 1979–1992. (Adapted from Madronich and deGruiji, 1992).

Latitude	% ozone loss 1979–1992	% increase in rate, basal cell carcinoma	% increase in rate, squamous cell carcinoma
55°N	7.4±1.3	13.5±5.3	25.4±10.3
35°N	4.8±1.4	8.6±4.0	16.0±7.6
15°N	1.5±1.1	2.7±2.4	4.8±4.4
15°S	1.9±1.3	3.6±2.6	6.5±4.8
35°S	4.0±1.6	8.1±3.6	14.9±6.8
55°S	9.0±1.5	20.4±7.4	39.3±15.1

As for malignant melanomas, these are far more dangerous, but so far the connection between these and UV-B exposure is not well understood. There is tentative evidence that UV-A, which is not absorbed by ozone, may be responsible, and that the cancer is triggered by short, intense exposures to UV radiation. If there is a connection, however, it would seem that the symptoms do not appear until long after the exposure has taken place.

There is understandably concern about the recent upsurge in melanomas, but before blaming ozone loss and increased UV levels at the ground, we should keep in mind that UV levels have not increased very much, at least not yet. In fact, the increase is so small it is roughly equivalent to moving around 50 kilometres nearer to the equator, equivalent to saying that moving from London to the south coast of England would trigger a sharp rise in cancer. Moreover, the upsurge began back in the 1940s, long before ozone depletion became a concern. The most likely explanation is the change in life-style most people in the western world have experienced during the second half of the twentieth century; many of us now take vacations at lower latitudes where there is naturally more UV radiation, and generally spend more time relaxing in the Sun. Neither should we lose sight of the fact that the incidence of cancers has been better reported in recent years, biasing the trends shown in Table 3.

Another concern is whether or not UV radiation can cause cataracts. Certainly, people living at low latitudes are far more prone to cataracts than those living in middle or high latitudes. However, as stated already, the increase in UV-B radiation over most of the planet is small, casting serious doubt on dramatic claims such as the one made in an article published by Newsweek in December 1991, stating that sheep in Chile are going blind in the spring because of cataracts induced by a rise in UV exposure. The Antarctic vortex is not frequently present over Chile (in some years, it doesn't travel that far north), and even when it is present, the additional UV reaching the ground is short-lived. The mostly glaring error is that severe ozone reduction has only been going on inside the Antarctic vortex since the mid 1980s, and in all that time it has spent perhaps a total of two months above Chile, nowhere near long enough to induce problems which take years to develop. In fact, the real cause of eye problems for the sheep (none of whom were even alive when ozone depletion began) has subsequently been shown to arise from infection.

A further concern about future changes in UV-B intensity is the impact this may have on plant and marine life. Such effects are proving very difficult to quantify, but the consensus of opinion is that many plants are expected to suffer deleterious effects. About half of the agricultural plants tested so far are sensitive to UV increases, although the severity varies from one species to another; some manage to adapt whilst others are seriously damaged. Moreover, the responses are highly variable even for plants of the same species.

As for marine life, many creatures live in the surface waters of the oceans which are remarkably transparent to UV radiation.

Whilst many of these animals do have built-in safeguards such as protective coatings, or bodies that actually work at night to repair damage done during the day, these natural mechanisms are believed to be triggered by changes in the intensity of *visible* light; the animals are therefore unlikely to respond to changes in the intensity of UV light.

We are fortunate, if one may call it that, to already have a kind of laboratory in which to study these effects. The waters around Antarctica lie beneath the ozone hole every spring, when more than 50% of the ozone is removed by heterogeneous processes in the stratosphere. At this time, UV levels at the ground increase by around a factor of three. However, less UV falls at high latitudes because of the long path through the atmosphere traversed by solar radiation, making the impact less severe than it would be for similar ozone loss at middle or tropical latitudes. One of the biggest dangers may be to phytoplankton, the microscopic plants which comprise the base of the marine food chain; phytoplankton are the staple diets of whales, for example. There has even been speculation that higher levels of UV-light might bring about a population collapse in marine phytoplankton, or at least reduce their photosynthetic production. Laboratory experiments have already demonstrated that increasing the amount of UV-A and UV-B inhibits phytoplankton photosynthesis. This is potentially serious since photosynthesis by plankton is far more efficient at absorbing carbon dioxide than the world's forests, the ratio being about 70:30%. Research suggests that the drop in phytoplankton productivity around Antarctica when the ozone hole is present is somewhere between 6 and 12%, but since the hole is only present for a quarter of the year, the annual effect is closer to 1.5–3%. Viewed against the natural year-to-year variability of 25%, this seems small, as is the contribution made by phytoplankton to the total carbon budget. At the same time, we should not lose sight of the fact that the above effect is set to increase as the ozone shield continues to thin in both polar regions and, to a somewhat lesser degree, in mid-latitudes.

Some argue that the damaging effects of excessive UV are already apparent; that there has been a decline in amphibian populations over the past decade. Whilst it is possible that increased UV-B has played a part, the most likely explanation, since the decline is world-wide and not confined to regions where UV-B has dramatically increased, is simply the stress placed on them by world-wide increases in human population.

(ii) Commercial Supersonic Aircraft and the Ozone Layer

The issue of how supersonic aircraft will affect the ozone layer was first debated back in the 1970s, by some of the same people who were also discussing the CFC-PSC issue. Once the CFC-PSC theory gained credence, however, attention shifted away from supersonic jets, in part because fleets of these aircraft were not scheduled for widespread production and use for at least several more decades. That time, however, is now much closer at hand, and once again concern over this potential hazard to ozone and climate has returned to the scientific and political arenas.

Supersonic aircraft, unlike the present day fleets of subsonic (slower than the speed of sound) aircraft, will spend most of their time in the stratosphere, their exhaust gases being injected directly into a region sensitive to catalytic ozone-destroying chemistry. The lowermost stratosphere is, in fact, something of an enigma: ozone profiles obtained from both ozonesondes and satellite-borne instruments suggest ozone loss rates are largest just a few kilometres above the tropopause (figure 13), the thermal barrier dividing the troposphere from the stratosphere (chapter 1.1). Even today, this is not particularly well understood, although the most likely mechanism is that air leaks out from the base of the stratospheric polar vortex during the winter, allowing chemically-processed air to reach sun-lit mid-latitudes where catalytic ozone destruction cycles can operate, therefore explaining the fact that the largest losses over mid-latitudes are observed in late winter.

In the stratosphere, vertical motion is slow and exhaust gases from supersonic aircraft will not easily be dispersed. In fact, they are expected to remain near the altitude of flight for several years. Not all of these exhaust gases are problematic, at least as far as ozone concentrations are concerned. The ones that really matter, according to current opinion, are the nitrogen compounds nitrogen oxide (NO), nitrogen dioxide (NO_2) and dinitrogen pentoxide (N_2O_5), collectively referred to as NOx, and the compound nitrous oxide (N_2O). The relationship between NOx and ozone is rather complicated, but the important point to appreciate is that in the troposphere, where subsonic aircraft fly, it actually *makes* ozone, whereas in the stratosphere, it destroys it, as does N_2O.

About two-thirds of the N_2O in the atmosphere is of natural origin, coming primarily from the decomposition of organic matter; the remainder is man-made. N_2O is actually quite un-reactive, and on average its molecules remain in the atmosphere for about 150 years. This allows plenty of time for it to reach the stratosphere where it is broken apart by ultra-violet radiation (much like CFCs). Supersonic aircraft, however, will be depositing N_2O directly into the stratosphere. The anticipated size of this fleet is about 500 aircraft (initially), each of which is expected to spend about 5 hours per day in the stratosphere. This would introduce roughly as much N_2O as is transported up from the Earth's surface by natural processes.

It was Professor Paul Crutzen, the joint 1995 Nobel laureate in chemistry (along with Sherwood Rowland and Mario Molina) who discovered, in 1969, that NO_x is a very efficient catalyst for the destruction of stratospheric ozone. He proposed the following chemical reactions:

$$NO + O_3 \Rightarrow NO_2 + O_2$$

and

$$NO_2 + O \Rightarrow NO + O_2$$

with the net effect that $O_3 + O \Rightarrow 2O_2$. In other words, these reactions would convert one molecule of ozone into two molecules of oxygen, provided the free oxygen atom (liberated from nitrogen dioxide) was present.

But recall that stratospheric NO_x chemistry is remarkably complicated, far worse than the chlorine chemistry discussed in chapter 2.2. NO_2, for example, has the ability to react with ClO, converting chlorine from this active form into the reservoir form $ClONO_2$, so that NO_x can both promote and suppress ozone loss at the same time. This is one of the reasons estimates of the effect that supersonic aircraft emissions will have on ozone have varied so wildly through the years, swinging from a net ozone loss to a net ozone gain, and back again.

The discovery of the ozone hole and the evolution of the science of heterogeneous chemistry has helped to shed some light on the processes involving aircraft emissions. In chapter 2.2, it was noted that the polar stratospheric clouds, which allow the conversion of chlorine from inactive to active forms, occur only at high latitudes where aircraft spend little of their time. However, research into the properties of PSCs has revealed that sulphate aerosols, globally present throughout the lower stratosphere, can also act as surfaces for heterogeneous reactions. Sulphate particles are, then, a medium on which N_2O_5 may be converted into nitric acid (HNO_3). It was shown in chapter 2.2 that this is a compound created during the process which activates chlorine.

The net effect of these reactions is to *reduce* the amount of NO_x present in the lower stratosphere, which makes the ozone-destroying catalytic cycles involving NO_x far less efficient than the one involving chlorine. With recent and ongoing improvements in supersonic aircraft design, their overall impact on stratospheric ozone may be small, reducing the total ozone layer by less than 1%. The future for the supersonic fleet looks promising, but there is still one more consideration, and that is the effect that the proposed fleet of supersonic aircraft and, perhaps more importantly, today's fleets of subsonic aircraft are expected to have on the climate system (chapter 3.4).

(iii) The Relationship Between Stratospheric Ozone Depletion and Climate Change

Whilst climate is, strictly speaking, the subject of Part Three of this book, there are causal connections between natural (and perhaps man-made) variations in climate and the future well-being of the ozone layer which merit discussion here. The phasing out of the chlorofluorocarbons family (CFCs), called for by the Montreal Protocol, has been successful in curbing the future destruction of the ozone layer. The chemical industry responded to the Protocol by developing alternative compounds with similar properties to CFCs, but which will be less harmful to the ozone layer. In fact, they began developing these alternatives whilst the legal battle still raged in the early 1980s, perhaps aware that their position was becoming increasingly untenable.

The first step in the process was a switch from CFCs to hydrochlorofluorocarbons (HCFCs), an improvement, but HCFCs could still destroy ozone, albeit less efficiently. These were followed by hydrofluorocarbons, HFCs, which have a zero ozone depletion potential, or ODP (Appendix 2.5) because they never reach the stratosphere. They are considerably less stable than their predecessors and break apart in response to longer wavelengths of UV radiation in the troposphere. At first glance, HFCs appear to be an ideal solution, but whilst harmless to ozone in a direct sense, they have enormous global warming potentials, or GWPs (Appendix 2.4). This means that they possess the ability to trap infra-red radiation reflected by the Earth's surface in the troposphere, acting as greenhouse gases and thereby warming the lower atmosphere. At the same time, this prevents the IR radiation from reaching the stratosphere, which consequently cools down. It seems paradoxical that the solution to one environmental problem which accords with the requirements of the Montreal Protocol, contravenes the requirements of another, the Kyoto Protocol (chapter 3.4), the first legally-binding document (as yet unratified) designed to control the enhanced greenhouse effect.

The connection between a changing climate and the ozone layer is subtle. For a number of years, there was a widespread misconception that the thinning ozone layer was solely responsible for the threat of climate change, but this is not the case at all. Certainly, ozone depletion is cooling the stratosphere, as is global warming, and this will have some impact on the climate, but its contribution is expected to be relatively small. The real causality lies in the opposite sense: a warm climate is likely to enhance the destruction of stratospheric ozone.

Before discussing how climate may influence the fate of the ozone layer, let us first consider what happens when the climate shifts naturally. We can gain valuable insights from this into the effects that changing the temperature will have on the amount of ozone in the stratosphere. From 1645 to 1715 A.D., during a period known as the Maunder Minimum, sunspot activity on the surface of the Sun ceased almost entirely. There is no direct evidence that the total irradiance of the Sun was reduced at this time, but the Sun was certainly behaving abnormally. The Maunder Minimum was actually part of a longer period of cooler climate, often referred to as the Little Ice Age, which was apparent from circa 1450 until 1800 A.D., the coldest decades coinciding with the decrease in sunspot activity. The mean surface temperature of the Earth during this period dropped by somewhere between 0.5° and 1°C. However, there were no meteorological records at that time, no balloon-borne instruments to probe the upper atmosphere, and no one alive who knew anything about gases. So how is it that we can know, in the absence of meteorological data, whether or not the ozone layer was affected by this small drop in temperature?

The only tools we have to probe this cold spell are computer -climate models. These allow us to simulate the conditions prevailing at the time of the Little Ice Age. Researchers at the Department of Atmospheric Sciences at the University of Illinois, under the guidance of Professor Donald Wuebbles, have used a two-dimensional model to recreate that now-vanished atmosphere, comparing their results to the conditions which existed at the beginning of the nineteenth century, when the Little Ice Age was ending and human activities had yet to make a significant impact on the composition of the atmosphere. Their model, like many others, includes dynamics (the transport of air by the winds, atmospheric waves, etc.), a radiative transfer scheme and some forty-eight applicable chemical reactions.

In Part One, it was stressed how different climate models are notorious for giving different results when simulating the same set of circumstances, so how could Wuebbles and his colleagues know that their model could accurately represent the real atmosphere of three or four hundred years ago? In truth, they couldn't know with absolute certainty; we probably never shall. But it is possible to be reasonably confident about the model's abilities if it can accurately reproduce the composition of today's atmosphere. We know how we have altered the concentrations of at least some of its trace constituents, which means we also know (to a close approximation) what the pre-industrial atmosphere must have been like. Wuebbles et alia therefore set about comparing 1990 ozone concentrations with those which we believe existed around 1800 A.D. The results of this test run are presented in figure 16A: here, the distribution of ozone depletion is seen as a vertical cross-section through the atmosphere as a function of latitude. The ordinate on the left shows atmospheric pressure decreasing from the ground upwards; on the right, the approximate pressure altitude derived from the hydrostatic equation in Appendix 1.2. The figure shows that the total column ozone in 1990 relative to circa 1800, at 5° intervals over the surface of the Earth and at 1.5 kilometre-intervals going up through the atmosphere (the resolution of the model) is severely depleted over the south polar regions, by almost exactly the right amount to accord with the measurements made of ozone depletion. It tells us, there was over 200% more ozone over Antarctica in 1880 than was observed in 1990.

In the Arctic, the model predicts that ozone has declined by more than 150% during this same period, which is again close to what we have witnessed. It also shows zero ozone loss in the tropics and only a little in mid-latitudes. With reasonable confidence in their model, the researchers set the concentrations of trace gases to levels present in the atmosphere around the year 1800, but lowered the temperature by 1°C to simulate the cooling during the Little Ice Age. The effects on the ozone layer are shown in figure

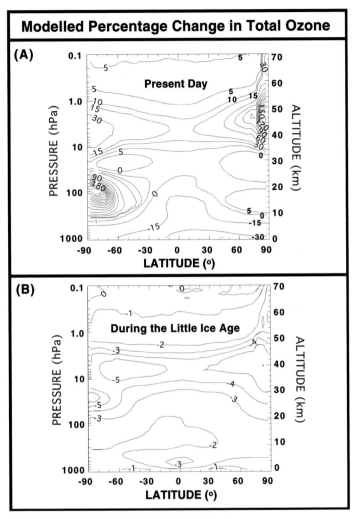

Figure 16 **A**, Present-day October concentrations of ozone are subtracted from modelled values, circa 1800 A.D., shown as a percentage change. Compared to the beginning of the nineteenth century, ozone has decreased dramatically in the polar regions of both hemispheres; **B**, modelled ozone concentrations during the Little Ice Age (which was coldest during the second half of the seventeenth century) are subtracted from the amount of ozone modelled for circa 1800 A.D. The negative values suggest that a colder atmosphere produces less ozone. This temperature effect is small compared to the magnitude of the anthropogenic changes in (A). (*Courtesy of Donald Wuebbles, University of Illinois*).

16B. The decrease in the total column is clearly a lot smaller, the largest only about 3.5%. There were no CFCs present in the atmosphere before 1930 (although sporadically there would have been volcanic aerosols), so this effect is largely a consequence of stratospheric cooling, almost certainly produced by a small drop in solar activity. This, of course, is something which could happen again at any time.

If the stratosphere were to cool down, then we might reasonably expect ozone depletion to worsen, even when levels of chlorine and bromine begin falling in the early years of the new millennium, as predicted. Many climatologists now agree that the lower stratosphere really is cooling down, and they expect this cooling to continue for at least a few more decades. Now recall how sensitive the circulation of the atmosphere is to changes in temperature (Part One), and that this is a positive feedback mechanism — a colder stratosphere promotes more efficient ozone-destroying chemistry, and as the amount of ozone decreases, less UV radiation is absorbed, allowing it to reach the Earth's surface. Since the absorption of UV by ozone is responsible for warming the stratosphere in the first place, the stratosphere will cool even more, enhancing the ozone-destroying chemistry.

The cooling of the stratosphere will, it is believed, affect ozone concentrations in a manner similar to that modelled for the Little Ice Age, but because there are now CFCs in the atmosphere, the result is likely to be far more severe. The destruction of ozone by heterogeneous chemistry is strongly temperature dependent; the colder the stratosphere, the more efficient these reactions become, a fact amply demonstrated over Antarctica where temperatures often fall below $-80°C$, equivalent to ~193 K, the threshold for the formation of type I PSC clouds.

In the northern hemisphere, the winter polar vortex now often persists well into spring, whereas just a decade ago it was usually gone by the end of February. A long-lived polar vortex in the northern hemisphere is not unique to the ozone loss era (in some winters in the 1960s, it was also very persistent), but there seems little doubt that stratospheric cooling is contributing to its longevity.

Until recently, it was widely accepted that the ozone layer would begin its century-long return to 1980 levels from around 2003, the time when chlorine and bromine concentrations in the stratosphere are expected to peak. However, this prediction does not take account of rising greenhouse gas emissions. Using a climate model once more, we can get an idea how far a cooling stratosphere is likely to delay the recovery of the ozone layer. The model in this case is one devised by NASA's Goddard Institute for Space Studies, or GISS.

Setting the parameters of the model for future conditions instead of those which prevailed in the past, as done by Wuebbles *et alia*, Drew Shindell and co-workers at GISS have calculated the probable changes in the ozone layer throughout the atmosphere. This model was also set the task, successfully, of reproducing today's observed polar ozone losses as a test of its validity. It was then run without a contribution from increased greenhouse gas (GHG) levels to establish a base-line for comparison with later runs, which would include elevated concentrations of GHGs. The result was a rise in temperature at the ground and a corresponding fall in temperature in the stratosphere. As found by the Illinois group, in the absence of halogen (chlorine and bromine) chemistry, a change of 1°C or so in temperature does not alter the amount of stratospheric ozone by more than a few percent. When halogens are present, however, the effect is quite dramatic; the model then predicts that the large ozone loss seen in the 1990s are likely to persist until around 2025, rather than 2003, both in the Antarctic and the Arctic. The most severe loss is expected to take place in the decade 2010–2019, when roughly 60% of the Antarctic ozone column is destroyed each spring (as now) and Arctic losses are above 50% over a large area. The earliest recovery to the kinds of values seen in the late 1980s (when ozone destruction was already becoming noticeable in the Antarctic) is not expected before around 2060. These results are presented in figure 17, the axes the same as for figure 16.

The upper four graphs show what is expected to happen during March in the northern hemisphere, decade by decade, until 2029; the bottom four are the same but for September in the southern hemisphere. In the Arctic, the most severe depletion occurs during the decade 2010 and 2019 between altitudes of 10 and 30 km. A reduction in ozone of up to 60%, almost as severe as the Antarctic loss in the 1980s, is expected between 70° and 75°N. Over Antarctica, losses will remain above 90% over a deep layer of the lower stratosphere, its vertical range maximising in the second decade of the twenty-first century. Even by 2029, the ozone in this region may barely have started to recover.

Quite apart from promoting additional loss of ozone via heterogeneous chemistry in a colder stratosphere, thereby delaying the

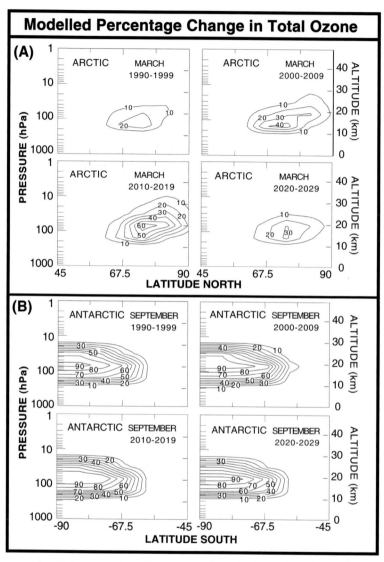

Figure 17 Future spring-time ozone loss in both hemispheres for the decades 1990–99, 2000–09, 2010–19 and 2020–29. The severity of ozone loss over Antarctica changes little during the total period, remaining at 90% in the lower stratosphere, whilst in the Arctic, the greatest loss is expected between 2010 and 2019, long after the chlorine loading in the stratosphere begins to decline. This delay in recovery is expected because the stratosphere is being cooled by increasing greenhouse gas concentrations. (*Courtesy of Drew Shindell, GISS*).

recovery of the ozone layer, an increase in greenhouse gases is also likely to alter the way in which ozone is transported from its source region in the tropical stratosphere to higher latitudes. Recall from Part One the causal relationship that exists between temperature and wind speeds, and hence with the atmospheric circulation. The GISS model suggests that the speed of the vertical circulation will increase, bringing more ozone down from higher altitudes in the polar regions where the ozone destruction cycle is highly effective. In contrast, the horizontal motion from equator to pole will probably slow down. In an atmosphere where the concentrations of carbon dioxide and other greenhouse gases are rising sharply, the way in which ozone is distributed in the stratosphere may change significantly in the future.

There is still one more aspect of the atmosphere's dynamical behaviour implied by the GISS model which offers cause for concern: the wave activity promoted by the numerous mountain ranges in the northern hemisphere may decrease in the Arctic if the circulation changes, and if this is the case, the circumpolar vortex will be disrupted less frequently than at present, allowing it to remain intact throughout the entire winter and facilitating still more ozone destruction.

If this model is even half-way right, the health of the ozone layer looks somewhat poor for much of the twenty-first century. This prediction is very recent, however (Nature, 1998), and has yet to be addressed in the constantly evolving Montreal Protocol devised to protect the ozone layer.

(iv) The Montreal Protocol

The Parties to the Protocol, being Parties to the 'Vienna Convention for the Protection of the Ozone Layer', mindful of their obligation under the Convention to take appropriate action to protect human health and the environment against adverse effects resulting, or likely to result, from human activities which modify, or are likely to modify, the ozone layer......

So began **"The Montreal Protocol on Substances That Deplete the Ozone Layer"**, concluded in its original form in Montreal on 16th September 1987. This document represented the first global agreement to initially restrict, and then ban altogether, the use of CFCs and other substances which threaten the ozone layer. This move was unprecedented insofar as it was the first time a significant fraction of the human race (at least, the 32 industrialised countries so designated at the time) had managed to agree on something with far-reaching implications for civilisation, apart from the policing of wars. Figure 18 shows how events unfolded, and are still unfolding, in connection with the Montreal Protocol.

The Protocol had its genesis in 1975, the year following the publication by Rowland and Molina warning of the danger of CFCs to the ozone layer. The World Meteorological Organisation (WMO) issued its first intergovernmental statement, also warning of these dangers, and in 1976 the United Nations Environmental Programme (UNEP) called for *"an examination of the need and justification for recommending any national and international controls over the release of man-made chemicals"*.

In the same year, the UNEP Governing Council authorised the convening of a meeting of experts assigned by their various governments to discuss what was then known about ozone. From this meeting emerged a *World Plan of Action* which involved monitoring atmospheric ozone, the amount of solar radiation reaching the Earth's surface, the impact of ozone depletion on human health and the ecosystem in general, and to develop ways to assess the cost and benefits of any control measures which may subsequently be introduced.

As scientific information steadily accumulated, so did concern about the future of ozone. In January 1982, UNEP convened the first meeting of *legal and technical experts* in the hope of further developing what was then only a Global Framework Convention for the protection of the ozone layer; in other words, a watchdog without teeth. It was not until March 1985 in Vienna, after three years of intense negotiations between the member countries, that a detailed scientific review formed the basis of a document comprising twenty-one articles, pledging to protect human health and the environment from the effects of ozone depletion. Plans were made for research into atmospheric ozone and for a free exchange of data and information, something hitherto rare in the scientific community. At this same meeting, the members also agreed to work towards preparing concrete measures in the form of a Protocol to the Convention, which transpired two years later.

In its original form, the Montreal Protocol called for a 50% reduction in the manufacture of CFCs by the year 2000, including

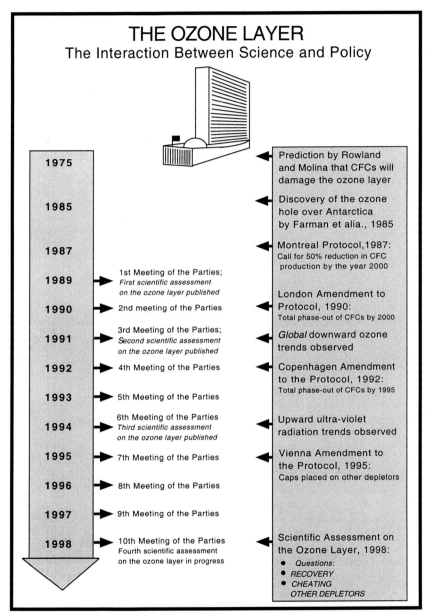

Figure 18 A time-line showing the interaction between scientific discoveries and policy decisions during the evolution of the ozone crisis, beginning from its first prediction in 1975. (*Courtesy of Daniel Albritton, NOAA Aeronomy Laboratory*).

Table 4 The phase-out schedule for chlorofluorocarbons called for by the Montreal Protocol and its subsequent amendments. (Adapted from the 1994 Scientific Assessment on Ozone Depletion).

1. CHEMICALS COVERED BY 1987 MONTREAL PROTOCOL

CFCs (11,12,113, 114,115)	Phase down 1986 levels by 20% by the end of 1994; 50% by the end of 1999.

2. THE MONTREAL PROTOCOL (LONDON AMENDMENT-1990)

CFCs (13,111,112, 211,212,213, 214, 215,216,217)	Phase down 1989 levels by: 20% 1993 85% 1997 100% 2000
Halons (1211,1301, 2402)	Freeze in 1992 at 1986 levels, then Phase down by: 50% 1995 100% 2000
Carbon tetrachloride	Phase down 1989 levels by: 85% 1995 100% 2000
Methyl chloroform	Freeze in 1993 Phase down 1989 levels by: 30% 1995 70% 2000 100% 2005

3. FURTHER STRENGTHENING OF THE MONTREAL PROTOCOL (COPENHAGEN AMENDMENT-1992)

CFCs	Phase out	100% by the end of 1995
Halons	Phase out	100% by the end of 1993
Carbon tetrachloride	Phase out	100% by the end of 1995
Methyl chloroform	Phase out	100% by the end of 1995
Methyl bromide	Freeze at 1991 levels	by the end of 1994
HCFs	Phase down 1989 levels:	35% by the end of 2004 90% by the end of 2014 99.5% by the end of 2019 100% by the end of 2029

a comprehensive schedule for the phasing-out of these substances (see Table 4).

In the years following this historic event, advances in scientific understanding made it clear that the original schedule for phasing out CFCs was inadequate; figure 19 indicates how the concentration of chlorine would increase in the stratosphere with the restrictions imposed by the original Montreal Protocol (curve A),

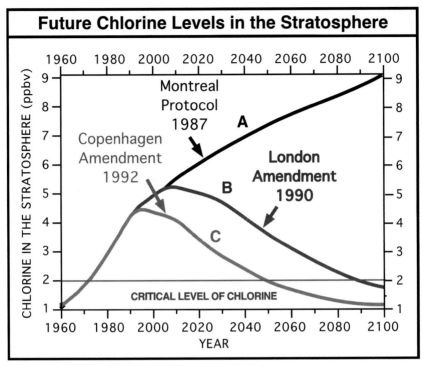

Figure 19 The time required for chlorine levels in the stratosphere to drop back below the critical threshold of 2 parts per billion, at which severe ozone destruction began. The first version of the Montreal Protocol did not call for sufficient reductions to prevent continued chlorine growth (curve A), but its subsequent amendments in London in 1990 (curve B) and in Copenhagen in 1992 (curve C) were both adequate to reverse this trend. Chlorine is now expected to fall below the critical threshold by around the year 2050.

and by the 1990 London Amendment (curve B). Even with a reduction of these compounds by the year 2000, chlorine levels were still going to increase dramatically for a hundred years, promoting colossal environmental damage.

A faster reduction and phase-out of CFCs was called for in the second amendment to the Protocol, at a meeting held in Copenhagen during 1992. This required that production be halted completely by no later than 1995. In this scenario, chlorine levels would then peak in the early years of the twenty-first century at about 3.5 ppbv, dropping back to below 2 ppbv by around 2050, at least 30 years earlier than called for by the 1990 amendment. The Copenhagen Amendment also required the phase out of the replacements for CFCs, hydrochlorofluorocarbons (HCFCs), which still have a limited capacity to destroy ozone, by no later than 2029. To facilitate this more rapid phase out of CFCs, an environmental fund was set up and paid for by the developed nations to ease the financial pressures.

All in all, then, the Montreal Protocol is a great success story. And yet, in spite of all the measures so carefully devised to protect the ozone layer, a problem remains. The CFC alternatives possess an even greater global warming potential (Appendix 3.1), and contribute in no small way to the *enhanced* greenhouse effect the subject of the next part of this book.

Appendices for Part Two

APPENDIX 2.1

MENDELÉEV'S PERIODIC TABLE OF THE ELEMENTS

There are 92 chemically distinct, naturally-occurring elements found on Earth. Despite the apparent diversity of elements, they are comprised of just *three* kinds of elementary particles, all of which were discovered by researchers at the University of Cambridge: the proton, the electron and the neutron.

The proton is a particle which carries a positive electrical charge, while the electron possesses a negative electrical charge, and because all atoms are electrically neutral they must contain protons and electrons in equal numbers. The neutron, as its name suggests, is electrically neutral. Protons and neutrons make up the nucleus of an atom, its *core*, around which circle electrons in a series of orbits, rather like the planets orbiting the Sun. This is the picture conceived by Niels Bohr (Appendix 2.3), and whilst not altogether accurate in the light of quantum physics, it is an edifying first approximation.

The first person to arrange the elements in a logical progression was the Swedish scientist Dimitri Mendeléev. In 1869, he constructed a table of the elements, called the Periodic Table (somewhat different to the one in common use today, figure 6), in which he arranged them in order of complexity. The first to appear in the table is hydrogen, the simplest and most abundant element in the Universe, making up more than 80% of all matter. Its nucleus contains a single proton (the number of protons is equal to an element's atomic number, in this case one) orbited by a single electron. The next element in order of complexity is helium which has two protons (atomic number 2) and two neutrons in its nucleus, orbited by two electrons. Next comes Lithium with three protons (atomic number 3), 4 neutrons and three electrons. The number of protons continues to increase by one for each successive element right up to Uranium with 92 protons (atomic number 92), ninety-two electrons and 144 or 146 neutrons.

The Periodic Table does more than simply arrange the elements in order of complexity; it also groups those which share certain characteristics. For example, consider the Alkali (non-acidic) elements. These appear in the columns labelled **A**; those with a single electron in their outermost electron shell all appear in column 1, and all those whose outermost electron shells are full appear in column 8. Having a full outer shell makes an atom very stable, which means its is unlikely to react with other atoms (or molecules); this group of elements is often referred to as the *Noble Gases*.

Column 7 of the table is shaded in grey to emphasise the fact that it contains all the ozone-destroying elements, collectively known as halogens. Each of these elements has a vacant space in its outermost electron shell (known to chemists as a 'free bond'), enabling them to combine readily with atoms such as oxygen (column A6 in the Periodic Table) which has two free bonds in its outermost shell. Moreover, the tendency for an atom to react with others is strongest at the top of the column, so that fluorine is more reactive that chlorine, which in its turn is more reactive than bromine, and so on.

Given that like electrical charges repel one another, it may seem extraordinary that the nucleus of an atom like Uranium (atomic number 92) can remain intact. It does so because of a nuclear force which acts over the very short distances between protons and is strong enough to overcome their mutual electrical repulsion. Neutrons also exert this force but, lacking an electrical charge, contribute nothing to the repulsion; they serve instead to keep a nucleus intact. Despite this attractive force, however, the nuclei of the heavier elements are not entirely stable, and they will often break apart spontaneously. This is the source of naturally-occurring radioactive decay.

Elements with atomic numbers greater than 92 (at present going up to atomic number 112) do not occur naturally and are all man-made.

APPENDIX 2.2

DEFINITION OF TOTAL OZONE AND THE DOBSON UNIT

Measuring the amount of ozone above each location on the Earth's surface came into use because this was the only quantity early ozone instruments were capable of measuring. Although theory did predict that most of the atmosphere's ozone lay in the stratosphere (indeed, *created* the stratosphere), this was not verified experimentally until the advent of balloon-borne instruments and rocket flights near the middle of the twentieth century.

The measurement of the total column of ozone is extremely useful since both ground-based and satellite instruments can measure this quantity, allowing us to monitor how much ozone is present at any given location without the need for balloons to traverse the atmosphere many times per day. (In fact, satellite-borne instruments are now capable of seeing the vertical structure of atmospheric ozone, at least down to as far as the tropopause).

The total amount of ozone above our heads, from the ground to the top of the atmosphere (the total ozone column) is defined as being equal to the number of ozone molecules, N_3, in a vertical column whose area at the ground is *one centimetre-squared* (1 cm^2), at a particular temperature and atmospheric pressure, known as standard temperature (0°C) and pressure (1000 hPa; sea level), or STP. The value of N_3 may be calculated using integral calculus:

$$N_3 = \int_{z=0}^{z=\infty} n_3 dz$$

where n_3 is the number density of ozone. If the atmosphere were reduced to STP, then the equivalent depth for total ozone would be

$$X_3 = \frac{N_3}{L_0}$$

where L_0 is Loschmidt's number = 2.69×10^{19} molecules cm^{-3}. For a column of air which contains 8×10^{18} ozone molecules cm^{-2} (the global average total amount of ozone per cm^2 above sea level), the equivalent depth for total ozone is 0.3 cm at STP, or 300 metre-atmosphere-centimetre (m-atm-cm). One m-atm-cm is a Dobson Unit, (DU, named after the Oxford meteorologist Gordon M. B. Dobson). One DU is 10^{-3} cm of ozone at STP, or 2.69×10^{16} molecules of ozone cm^{-2}. Thus, at STP a single Dobson Unit would stand just 0.01 mm high!

In the tropics, there are around 250 DU throughout the vertical depth of the atmosphere, whilst in the northern hemisphere polar latitudes this value is 450–500 DU. Southern hemisphere values used to be around 400–450 DU, naturally lower than in the northern hemisphere, but since the advent of the ozone hole this value has fallen below 100 DU during spring.

The Earth's surface area is approximately 1.72×10^{14} m^2, which means there are something like 4×10^{37} ozone molecules in the atmosphere. The envelope of air surrounding our world is comprised of about 10^{44} molecules of all types, roughly ten million times bigger than the number of ozone molecules, emphasising that ozone really is just a *trace* gas.

APPENDIX 2.3

THE PLANCK CONSTANT, h

When considering interactions between atoms and molecules, it is not entirely appropriate to think in terms of the energy which powers these interactions being continuous in nature. Quantum physics, the investigation of the sub-atomic world, allows that light may travel not only in the form of waves which have different lengths and frequencies, but also, and simultaneously, as discrete packets of energy called quanta. Within atoms, electrons often move from one energy level to another, or one orbit to another in the framework of classical physics. However, a change in level only occurs in response to discrete quanta; too much or too little energy will evoke no response. The equation which governs the absorption (and emission) of radiation of all kinds is

$$\mathbf{E} = \mathbf{h\nu}$$

where $\mathbf{E}$ (unit: *joules*) is the energy and ν the frequency of the light (unit: *hertz*). The constant of proportionality, $\mathbf{h}$, is called Planck's constant, after Max Planck who introduced it in the year 1900. It's value is 6.626176×10^{-34}, in units of *joule seconds*.

Because electromagnetic radiation travels at the speed of light, c, we may relate frequency and wavelength using the equation

$$\nu = \frac{c}{\lambda}$$

where λ is the wavelength (unit: metres). It follows from this that the amount of energy contained in a photon of radiation is inversely proportional to the wavelength of that radiation.

APPENDIX 2.4

THE CHEMISTRY OF OZONE PRODUCTION AND LOSS

In 1928, Gordon Dobson identified the region between ~12 km and 50 km as being the seat of the Earth's ozone layer. Two years later, Sidney Chapman proposed a set of four chemical reactions to account for the formation of ozone in the stratosphere:

$$O_2 + hv \Rightarrow O + O \qquad [1]$$
$$O + O_2 + M \Rightarrow O_3 + M \qquad [2]$$
$$O_3 + hv \Rightarrow O + O_2 \qquad [3]$$
$$O + O_3 \Rightarrow O_2 + O_2 \qquad [4]$$

Reactions [2] and [3] recycle odd oxygen (O_x) rapidly whilst [1] and [4] create and destroy O_x comparatively slowly. It is therefore reactions [1] and [4] which govern the rate at which ozone is created. We know today that catalytic processes accelerate reaction [4]:

$$X + O_3 \Rightarrow XO + O_2 \qquad [5]$$
$$XO + O \Rightarrow X + O_2 \qquad [6]$$

with the net effect that

$$O + O_3 \Rightarrow 2O_2 \qquad [7]$$

The destruction rate of ozone depends both on the reaction rates *and* the concentrations of the catalysts.

The use of 'X' in the above equations suggests that a number of chemical compounds may fulfil this role. These compounds include the families of odd-nitrogen (NO_x), odd-chlorine (Cl_x) and odd-hydrogen (HO_x). The odd-nitrogen family contains the elements or compounds atomic nitrogen (N), nitrogen monoxide (NO) and nitrogen dioxide (NO_2). The main source of NO_x

production in the stratosphere arising from the photolysis of nitrous oxide (N_2O).

The odd-hydrogen family contains the species atomic hydrogen (H), hydroxyl (OH) and hydroperoxyl (HO_2). In the lower stratosphere, the hydroxyl and hydroperoxyl radicals react catalytically to destroy ozone via the reactions:

$$OH + O_3 \Rightarrow HO_2 + O_2 \qquad [8]$$
$$HO_2 + O_3 \Rightarrow OH + 2O_2 \qquad [9]$$

with the net effect that

$$2O_3 \Rightarrow 3O_2 \qquad [10]$$

The hydroxyl radical is formed through the reaction of excited atomic oxygen (O^{1D}) with water vapour, molecular hydrogen (H_2) and methane (CH_4). OH also reacts with NO_2 to produce nitric acid (HNO_3):

$$OH + NO_2 + M \Rightarrow HNO_3 + M \qquad [11]$$

where M is a minister molecule, a compound which is itself unchanged by the reaction but which assists in the necessary exchange of energy involved. Nitric acid is a reservoir for NO_x, and its long lifetime allows its removal from the stratosphere by being rained out in the troposphere. Odd-hydrogen is also lost via reactions with nitric acid and methane:

$$OH + HNO_3 \Rightarrow H_2O + NO_3 \qquad [12]$$
$$CH_4 + OH \Rightarrow CH_3 + H_2O \qquad [13]$$

The odd-chlorine family of compounds which catalytically destroy ozone in the stratosphere consists of Cl, ClO, HCl, $ClONO_2$ and HOCl. The primary sources of Cl_x are methyl chloride (CH_3Cl), originating in the oceans, and chlorofluorocarbons (CFCs) whose sources are man-made. In the stratosphere, Cl_x radicals are produced from the photodissociation of CFCs and CH_3Cl, although OH radicals provide a tropospheric sink for the latter; hence, only a small fraction of emitted CH_3Cl reaches the stratosphere. The catalytic destruction of ozone by Cl_x radicals occurs chiefly in the middle and upper stratosphere via reactions [5] and [6], where Cl replaces 'X'.

The sinks for Cl_x are hydrochloric acid (HCl) and chlorine nitrate ($ClONO_2$), although chlorine radicals complete many cycles of destroying odd-oxygen before being converted to HCl via

$$Cl + CH_4 \Rightarrow HCl + CH_3 \qquad [14]$$

As with nitric acid, HCl is eventually rained out in the troposphere, although some chlorine atoms can be re-cycled through the reaction:

$$HCl + OH \Rightarrow Cl + H_2O \qquad [15]$$

Away from the polar regions, the three catalytic cycles of odd-nitrogen, odd-hydrogen and odd-chlorine all compete for ozone destruction in the altitude region 15–25 km, whereas between 25 and 40 km, odd-oxygen destruction is dominated by the reactions involving nitrogen. In the troposphere, odd-oxygen destruction occurs through the reaction of excited odd-oxygen (O^{1D}) with water, and ozone with OH.

In the polar regions, particularly in the southern hemisphere, ozone destruction is affected by heterogeneous chemistry. During the polar night, when no sunlight is present for months, photolysis is impossible and no ozone is produced. When temperatures are sufficiently low ($\leq 193K$), polar stratospheric clouds (PSCs) tend to form and provide surfaces on which heterogeneous chemical reactions (that is, reactions involving more than one phase) may take place.

There are two types of PSCs, and their formation is highly temperature dependent. Type I PSCs which consist of nitric acid trihydrate (NAT; $HNO_3/3H_2O$) form at 193K (–80°C), whereas type II PSCs (comprised of cirrus-like clouds of water-ice particles) form by condensing on small NAT particles or other aerosols at 188 K (–85°C), provided at least 5 ppmv of water are present. On the surfaces of these PSCs, the relatively long-lived reservoir species of chlorine, HCl and $ClONO_2$, can react to liberate chlorine into its gas-phase state:

$$HCl + ClONO_2 \Rightarrow Cl_2 + HNO_3 \qquad [16]$$

As the amount of gas-phase chlorine increases, so catalytic ozone destruction accelerates when sunlight returns in spring. Associated with PSC formation is denitrification, the removal of HNO_3

by sedimentation which prolongs the existence of elevated levels of gas-phase chlorine.

Ozone destruction proceeds via the ClO dimer (Cl_2O_2) catalytic cycle:

$$2(Cl + O_3 \Rightarrow ClO + O_2) \qquad [17]$$
$$ClO + ClO + M \Rightarrow Cl_2O_2 + M \qquad [18]$$
$$Cl_2O_2 + h\upsilon \Rightarrow OClO + Cl \qquad [19]$$
$$OClO + M \Rightarrow Cl + O_2 \qquad [20]$$

with the net effect that

$$2O_3 + h\upsilon \Rightarrow 3O_2 \qquad [21]$$

A further reaction,

$$BrO + ClO \Rightarrow BrO_2 + Cl \qquad [22]$$

provides an additional recycling mechanism for odd-chlorine.

The above reactions are important because the formation and photolysis of the ClO dimer molecule causes an ozone loss rate proportional to the square of the ClO abundance, and a single atom of chlorine can destroy hundreds or thousands of ozone molecules before being permanently removed from the stratosphere.

APPENDIX 2.5

OZONE DEPLETION POTENTIAL

The ozone depletion potential (ODP) of a compound is a simple measure of its ability to destroy stratospheric ozone. It is a relative measure: the ODP of CFC-11 is defined to be 1.0, and it is calculated for other compounds relative to CFC-11, so that a compound with an ODP of 0.1 is only 10% as effective at destroying ozone as CFC-11.

Mathematically, the ODP of a compound 'X' is defined as the ratio of the total amount of ozone destroyed by a fixed amount of compound 'X' to the amount of ozone destroyed by the same mass of CFC-11:

$$ODP(X) = \frac{-\Delta O3(X)}{-\Delta O3(CFC-11)}$$

The right-hand side of the equation is calculated by combining information from laboratory and field measurements with atmospheric chemistry and transport models. This method of assessing ODP is quite robust, for although there are many uncertainties on the right-hand side of the equation, these mostly cancel out when the ratio is calculated. Bromine-containing halocarbons generally have much larger ODPs than chlorocarbons because bromine is a far more effective ozone-destruction catalyst than chlorine.

The ODP as defined above is a steady-state or long-term property. As such, it can be misleading when considering the possible effects of CFC replacements. Many of the proposed replacements have short atmospheric lifetimes; however, if a compound has a short stratospheric lifetime, it will release its chlorine or bromine atoms sooner than a compound with a longer stratospheric lifetime. Therefore the short term effect of such a compound on the ozone layer is larger than would be predicted from the ODP alone.

To circumvent the above shortcoming with the ODP definition, a time-dependent ODP was introduced by Susan Solomon and Daniel Albritton of the NOAA Aeronomy Laboratory (reported in the World Meteorological Organisation's "Assessment of the Ozone Layer, 1991"). This defines ODP as follows:

$$ODP(X,t) = \frac{\text{Loss of ozone due to compound X in time t}}{\text{Loss of ozone due to CFC-11 in time t}}$$

BIBLIOGRAPHY FOR PART TWO

Most of the following list of books and articles are easily accessible, whilst the more specialised material appears chiefly in scientific journals. These tend to be rather expensive and are aimed specifically at the scientific reader (hardly surprising since they are also written by scientists!). If you wish to delve into this literature, the best way is through your local library. If this fails, your nearest University library will allow you access to it.

A very complete picture of the ozone issue is provided by the series of ozone assessment books. These contain contributions by dozens of atmospheric scientists around the world. They are designed (to some extent) to be read and understood by policy makers, but they still assume a considerable amount of specialist knowledge. References are also included for these, but they are not available in bookshops. To obtain them, write to the following address:

World Meteorological Organisation,
Global Ozone Observing System,
41 Avenue Giuseppe Motta,
P.O. Box 2300, Geneva 2, CH 1211,
Switzerland.

These volumes used to be free upon request, but there may now a small fee. You will also be expected to pay for postage and packaging.

BOOKS

Roan, S. L. "Ozone Crisis". Published 1989.
This book takes an in-depth look at the machinations between science, industry and politics in the years leading up to the discovery of the ozone hole. This enlightening insight is a glimpse

of what probably awaits us as the issues surrounding climate change gather momentum.

Scientific Assessment of Ozone Depletion: 1989. World Meteorological Organisation Global Ozone Research and Monitoring Project Report #20.

Written and published in the years immediately following the discovery of the ozone hole, some of the scientific opinions in this volume are now out of date. Its value, however, is that it contains a lot of introductory information which is not repeated in later reports. Read in conjunction with a later issue (see below), it also serve to demonstrate the changes in scientific opinion even on time-scales of a few years. It is almost 500 pages long, so be prepared to expend some effort.

Scientific Assessment of Ozone Depletion: 1994. World Meteorological Organisation Global Ozone Research and Monitoring ProjectReport # 37.

This volume is almost state-of-the-art, but will be superseded by the 1998/99 issue which, at the time of writing, is still being compiled.

PUBLICATIONS IN SCIENTIFIC JOURNALS

Dobson, G. M. B. "Forty years' research on atmospheric ozone at Oxford". *Applied Optics*, volume **7**, page 387, 1968.
This article is a nice summary of the evolution of our understanding of atmospheric ozone, by one of the pioneers in the field. Most of the ideas discussed herein are still valid today.

Farman, J. C., B. G. Gardiner and J. D. Shanklin. "Large losses of total ozone in Antarctica reveal seasonal ClO_x/NO_x interaction". *Nature*, volume **315**, page 207, 1985.
This is the scientific paper which announced the discovery of the ozone hole.

McCormick, M. P., H. M. Steele, P. Hamill, W. P. Chu and T. J. Swissler. "Polar Stratospheric Cloud Sightings by SAM II". *Journal of Atmospheric Science*, volume **39**, page 1387, 1982.

Molina, M. J. and F. Sherwood Rowland. "Stratospheric sink for chlorofluorocarbons: Chlorine – atom-catalyzed destruction of ozone". *Nature*, volume **249**, page 810, 1974.
The ground-breaking paper which first brought to light the potential hazard of CFC emissions to the ozone layer. Not for the faint-hearted!

Shindell, D. T., D. Rind and P. Lonergan. "Increased polar stratospheric ozone loss and delayed eventual recovery owning to greenhouse-gas concentrations". *Nature*, volume **392**, page 589, 1998.
This is the first publication concerning the delayed recovery of the ozone layer.

Solomon, S. "Progress towards a quantitative understanding of Antarctic ozone depletion". *Nature*, volume **347**, 1990.
This article is fairly general and was written with the non-specialist in mind.

Stolarski, R. S., P. Bloomfield, R. D. McPeters and J. R. Herman. "Total ozone trends deduced from Nimbus-7 TOMS data". *Journal of Geophysical Research Letters*, volume **18**, page 1015, 1991.

Trenberth, J. E. "Recent observed interdecadal climate change in the Northern Hemisphere". *Bulletin of the American Meteorological Society*, volume **71**, pages 988–993.
This article is rather advanced for the casual reader, but the author's style is lucid and informative.

Tuck, A. F. "Synoptic and Chemical Evolution of the Antarctic Vortex in the late winter and spring, 1987: An ozone processor". *Journal of Geophysical Research*, volume **94**, page 11687, 1989.
The first scientific paper suggesting that air flows through the Antarctic polar vortex instead of remaining isolated within it throughout the winter. The work remains contentious, but has strong implications for mid-latitude ozone loss.

Part Three
Climate Change

CHAPTER 3.1

CLIMATE AND THE GREENHOUSE EFFECT

What exactly is meant by *climate*? The term is often used loosely and there is, understandably, some confusion about what scientists intend when they speak of *climate change*. The climate, after all, is changing from day to day on the local scale, and from one epoch to the next as ice ages wax and wane.

From a scientific perspective, *climate* is taken to mean the *average physical environment of the Earth*, as defined by temperature, rainfall, humidity, wind speeds and directions, the amount of incoming solar energy and variations in atmospheric pressure. Broad zones of different climate exist on the Earth: the tropics (the region between around 30°N and 30°S contained by the subtropical jet streams described in chapter 1.2); the temperate mid-latitude bands where most of us live, and the polar regions, the area enclosed by the polar circle. Unfortunately, there is more than one climate classification system in use by the scientific community, although most of them use a combination of the above factors, together with topography and soil content, to define climatic zones.

Fundamentally, climate is controlled by the long-term balance of energy in the Earth-atmosphere system, which means that, in the long run, the amount of incoming radiation from the Sun must balance exactly with the amount of radiation the Earth reflects back into space to maintain a stable climate system. Whilst the Earth absorbs energy mostly in the visible range of the electro-magnetic spectrum (Appendices 1.4 and 1.7), it re-radiates it chiefly in the form of infra-red radiation, so that the outgoing wavelengths are longer than those incoming. This is an important aspect of the greenhouse effect.

The natural variability in the climate system is very large, changing slowly over time according not only to shifts in the Earth's orbit about the Sun, but also to fluctuations in the activity of the Sun itself. The most recent example is the Little Ice Age. Other natural events which affect the climate, in the short term, are large volcanic eruptions such as that which occurred in the Philippines in 1991, when Mount Pinatubo injected vast quantities of sulphur and other solid material (aerosols) into the stratosphere. The aerosols from Pinatubo encircled the world in a matter of weeks and spread to high latitudes after just a few months, remaining aloft for several years and providing billions of particles on which heterogeneous chemistry could proceed and attack the ozone layer. Airborne material from large-scale eruptions also screens out more of the Sun's radiation which, in the short term, temporarily cools the Earth's surface, inducing temporary, negative climate forcing (Appendix 3.2).

Human activities are also capable of affecting the climate by changing the composition of the atmosphere, adjusting the energy balance and altering its circulation patterns. Because the Earth's oceans and atmosphere are intimately linked, circulation patterns in the oceans may also change. The consequences of this are far reaching because the oceans act as a gigantic energy reservoir, changing their characteristics far more slowly than the atmosphere (chapter 3.3)

A widely discussed component of the climate system is the *greenhouse effect*, mentioned above. The Earth absorbs all but 30% of incoming electromagnetic radiation, the difference being reflected directly back into space by clouds and therefore not interacting with the climate system at all. The 70% reaching the ground is absorbed and re-radiated at longer wavelengths – the light, changed from visible to infra-red radiation, now presents a larger cross-section to the atmosphere. These waves are intercepted by certain gases (and aerosols), and when these gases absorb radiation, they effectively trap energy in the lower atmosphere. This energy is then released again and is subsequently re-absorbed many times, no longer able to escape to space. This is the so-called *greenhouse effect*, and the atmospheric gases which contribute to it are called *greenhouse gases*. Without them, the Earth would be approximately 32°C cooler than it is. Greenhouse gases (GHGs) occur only in trace amounts, but it is clear from data spanning past glacial periods that altering their concentrations can have quite a dramatic effect.

The principal, naturally-occurring GHGs are listed in Table 5, indicating how their atmospheric concentrations have changed over the past two centuries. Water vapour, by far the most abundant greenhouse gas and by no means a tracer, is not listed because there is so much uncertainty about its contribution to the greenhouse effect. In a warmer world, more water would be present in the form of water vapour, some of which would form clouds and reflect back solar radiation, causing the atmosphere to cool. But if water vapour levels increase, they will alter not only the temperature of the atmosphere but also global weather patterns. Thus, the sense in which an increase in the amount of water vapour and cloud cover would affect the climate is proving very difficult to determine.

Although water vapour is the principal greenhouse gas in the atmosphere, this does not mean that the contributions of trace gases, including those manufactured by mankind, are insignificant. The most important of the *trace* greenhouse gases, in terms of global warming potential, or GWP (Appendix 3.1) is carbon dioxide (CO_2). Table 5 shows how the concentration of CO_2 has increased from around 280 parts per million by volume (ppmv) in the late 18th century to 361 ppmv in 1997. Its concentration is currently growing at around 1.6 ppmv every year, and is expected to double its pre-industrial level before the end of the 21st century. The primary reasons for its sharp increase are the widespread combustion of fossil fuels and the production of cement. Changes in land use have also made a significant contribution as forests are cleared to make way for agriculture and cities. In the thousand years prior to 1800 A.D., atmospheric CO_2 levels changed by less than 10 ppmv, and even then the change was gradual; an increase of this magnitude now occurs every six years. If this trend continues, CO_2 concentrations in the atmosphere will reach 750 ppmv towards the end of the twenty-second century (sooner if the emission rate increases). Our saving grace may be that we shall run out of fossil fuels before this time, and that such levels will never be achieved.

The next greenhouse gas, in order of GWP, is methane (CH_4). In fact, CH_4 presents us with something of a conundrum because its rate of growth is proving to be highly variable: during the 1970s it was increasing annually at a rate of 20 parts per billion by volume (ppbv), but in the 1980s this dropped back inexplicably to 9–13 ppbv before stopping altogether for a short while during 1992. Subsequently, it began to increase again. A major concern about rising methane concentrations in the atmosphere is the ability of this gas to affect the concentrations of other compounds which help to regulate the removal of CH_4. If these compounds themselves are decreased or removed, then methane levels may rise far more rapidly.

Table 5 List of the greenhouse gases (natural and man-made).

Species	Symbol	Lifetime (years)	Atmospheric concentration (ppbv)	Pre-industrial concentration (ppbv)	Annual rate of growth (ppbv)
Natural GHGs					
Carbon dioxide	CO_2	Variable	361,000	278,000	1,600
Methane	CH_4	12.2	1,714	700	8
Nitrous oxide	N_2O	120	311	275	0.8
Methyl chloride	CH_3Cl	1.5	0.6	0.6	0
Methyl bromide	CH_3Br	1.2	0.01	<0.01	0
Chloroform	$CHCl_3$	0.51	0.012	Unknown	0
Methylene chloride	CH_2Cl_2	0.46	0.03	Unknown	0
Carbon monoxide	CO	0.25	50–100	Unknown	0
CFCs					
CFC-11	CCl_3F	50	0.268	0	0
CFC-12	CCl_2F_2	102	0.503	0	0.007
CFC-113	CCl_2FCClF_2	85	0.082	0	0
CFC-114	$CClF_2CClF_2$	300	0.020	0	?
CFC-115	CF_3CClF_2	1,700	<0.01	0	?
Carbon tetrachloride	CCl_4	42	0.132	0	−0.0005
Methyl chloroform	CH_3CCl_3	4.9	0.135	0	−0.01
Halon-1211	$CBrClF_2$	20	0.007	0	0.00015
Halon-1301	$CBrF_3$	65	0.003	0	0.0002
Halon-2402	$CBrF_2CBrF_2$	20	0.0007	0	?
HCFCs					
HCFC-22	$CHClF_2$	12.1	0.1	0	0.005
HCFC-141b	CH_3CFCl_2	1.4	0.002	0	0.001
HCFC-142b	CH_3CF_2Cl	6.1	0.006	0	0.001
Perfluorinated compounds					
Sulphur hexafluoride	SF_6	3,200	0.032	0	0.0002
Perfluoromethane	$CF4$	50,000	0.07	0	0.0012
Perfluoroethane	$C2_{26}$	10,000	0.004	0	?
Some HFCs					
HFC-23	CHF_3	264		0	?
HFC-32	CH_2F_2	5.6		0	?
HFC-41	CH_3F	3.7		0	?
HFC-43-10mee	$C5H_2F_{10}$	17.1		0	?
HFC-125	C_2HF_5	32.6		0	?
HFC-134	CF_2HCF_2H	10.6		0	?
HFC-134a	CH_2FCF_3	14.6		0	?
HFC-143	$CF2HCH2F$	3.8		0	?
HFC-143a	CH_3CF_3	48.3		0	?
HFC-152a	CH_3CHF_2	1.5		0	?

Next comes nitrous oxide (N_2O). Levels of N_2O are rising in the atmosphere because there exists an imbalance between sources and sinks (or generation and removal) of this gas. At the moment, natural sources are thought to be around twice as great as anthropogenic ones, but the balance is slowly shifting in favour of the latter.

There are numerous other naturally occurring gases, but their contributions are, at present, very small. In addition to the naturally-occurring greenhouse gases are those we ourselves have manufactured. There are a great many compounds which were not present in the atmosphere before this century. ChloroFluoro-Carbons, or CFCs (discussed in Part Two because of their dramatic role in ozone destruction), together with their replacements, the HCFC and HFC families of compounds, are all powerful greenhouse gases. These compounds are discussed at greater length in chapter 3.4 in connection with future climate scenarios.

A critical point to appreciate about the greenhouse effect is that it is not something man-made. It has played a rôle in maintaining a habitable world for several billions of years, and is what keeps the mean surface temperature at a fairly constant 14°C. Computations show that, without the greenhouse effect, the mean surface temperature would be around −18°C, a little cold for most forms of life. Its influence only becomes problematic when this natural effect is augmented. An often-cited example of a runaway greenhouse effect is the state of the atmosphere on our sister world, Venus. The planet is almost identical to the Earth in both size and mass, and is not so close to the Sun that its atmosphere should be markedly different from ours. And yet, at some time in its history large quantities of carbon dioxide leaked into its atmosphere, so that now that atmosphere is 90% CO_2. In consequence, its surface temperature, as measured by space probes, is 477°C; without the greenhouse effect, it would be below −18°C, as cool as (if not cooler than) the Earth despite being closer to the Sun.

Incidentally, the value of 14°C for the mean surface temperature of the Earth has been obtained by averaging the temperature record between 1950 and 1970. But why just for this interval of time? Why not for longer? After all, the larger the quantity of data, the smaller the errors. The reason is first of all to compare the near-present-time temperatures with those of the past, and secondly (and this is why the averaging stops at 1970), to demonstrate the climb in global temperatures during the last 30 years of the twentieth century.

The idea of an augmented greenhouse effect is nothing new; in 1861, a scientist named Tyndall postulated that changes in the concentrations of greenhouse gases, such as water vapour and carbon dioxide, would warm the world, whilst at the end of the nineteenth century the Swedish chemist Svante Arrhenius wrote about it as something which might actually be beneficial to mankind. This view is highly simplistic of course, ignoring the complex interactions between atmosphere, oceans and land masses unknown in Arrhenius's time. It also ignores the effects of changing the radiative forcing of the Earth-ocean-atmosphere system.

Radiative forcing (also called climate forcing; Appendix 3.2) is the specific name given to the energy balance between incoming and outgoing solar radiation in the Earth-atmosphere system, and when scientists speak of a change in radiative forcing, they refer to a change in this energy balance, measured in units of energy (the watt, **W**, the same as the electrical supply in your home) per square metre of the Earth's surface: Wm^{-2}. The determination of the climate's response to a change in radiative forcing is complicated by the presence of feedback mechanisms, some of which can amplify warming effects (positive feedback) and others which reduce then (negative feedbacks). An increase in the amount of water vapour in the atmosphere, for example, which may happen as the oceans warm up and store heat on time-scales of hundreds to thousands of years (chapter 3.3), may serve to greatly enhance the greenhouse effect in the future. A negative feedback may also arise, however, since an increase in water vapour will increase the percentage of cloud cover over the Earth, which would reflect back more radiation into space without ever interacting with the climate system. This effect does depend greatly on the type and altitude of the clouds formed, which, as mentioned earlier, makes it extremely difficult to predict how large an impact a change in cloud cover would have on climate.

Before discussing the changes which have been observed in the climate system in recent times, we should first address how the climate has varied in the past, placing the recent rise in mean global temperature of 0.5–0.6°C into perspective.

SUN AND CLIMATE: A MESSAGE IN THE ICE

Thirty millennia ago, before mankind began living in settlements and farming the land, the Earth was a very different place. Our ancestors a thousand generations removed knew a world bounded to the north by glaciers stretching from the pole as far south as Britain, the middle of the United States and of Russia. For more than a hundred thousand years, all of Scotland and half of England were buried under one and a half kilometres of ice. The big thaw came as recently as ten thousand years ago when the last ice age ended.

This ice age was merely the last of many which have swept across the world during the past three million years. Although humans were present on Earth throughout this entire period, no one had yet started recording the events in the world around him. Man's attempts to record information for posterity date back less than 30,000 years. How is it, then, that we know about the temperatures and composition of the ancient, pre-historic atmosphere? How did we unlock the secrets of past climate?

A temperature record dating back millions of years has been deduced from the changing proportions of heavy to light atoms of oxygen in water. The technique is similar to carbon dating which utilises the ratio of carbon-14 to carbon-12 in living organisms such as trees. These are *isotopes* of the same element: this means that the number of protons is the same in each case (otherwise it would not still be carbon), but the numbers of neutrons present in the nucleus differ.

Oxygen and hydrogen are the constituent elements which form water (H_2O). Hydrogen can exist with either a single proton in its nucleus (regular hydrogen) or a proton plus a neutron (this form is called *heavy* hydrogen). Similarly, oxygen can exist as oxygen-17 (eight protons and 9 neutrons in the nucleus) or oxygen-18 (eight protons and 10 neutrons in the nucleus). By examining the ratio of the isotopes of oxygen in the water molecules comprising the oceans and the polar ice sheets, using an instrument called a mass spectrometer (in which a magnet sorts electrically-charged molecules according to their weights), it is possible to deduce information about past temperatures.

The longest climate records based on the isotopic composition of sea water come from the sea floor, where oxygen atoms in the calcium carbonate (chalk) of sea shells mirror the isotopic ratio in the sea when the shells were growing. The relation between these isotopes and temperature derives from the tendency of the heavier molecules of water in the ocean to evaporate more slowly than the lighter ones, and for the heavier molecules to condense more readily from the vapour state to the liquid state. In colder air, these heavier molecules are even less likely to exist in the vapour form, so that fewer heavy water molecules are found in snow which has been compressed to form the polar ice sheets. In fact, the pattern of winter and summer is as distinct in the ice as it is in the growth rings of a tree. Because this temperature information derives from an indirect method, rather than a direct measurement (such as measuring the temperature of the air right now with a thermometer), it is known as a *proxy* record. Thus, both ocean sediment and the ice in the polar caps provide a record of past temperatures.

But what about the composition of the ancient atmosphere? How can this also be known? This information also comes from polar ice, for not only does drilling down thousands of metres into the ice sheets of Greenland and Antarctica disclose the tree-ring-like pattern of ordinary and heavy water, it also brings to the surface bubbles of air which have been trapped in that ice for many millennia. In effect, tiny pockets of the ancient atmosphere have been frozen in time, the air inside of them containing the same concentrations of gases, such as carbon dioxide and methane, as when they were formed.

The deeper the position of the drill in the ice, the farther back into time the record goes. At Vostok Station in Antarctica, Russian and French scientists have recently drilled down to depth greater than two kilometres to recover ice deposited 160,000 years ago, during the last ice age but one. Figure 20(*A-C*) shows how temperature, and the concentrations of carbon dioxide and methane, have changed during this period.

Temperature is presented as deviations from today's mean surface temperature (14°C), which enables you to see, relatively, how much colder the world is during an ice age. This turns out to be a lot less than one might expect: in the depths of an ice age,

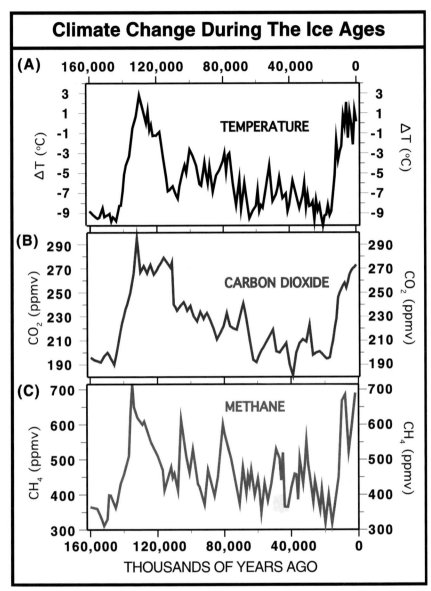

Figure 20 Relative temperature data (A), plus carbon dioxide (B) and methane (C) concentrations over the past 160,000 years, derived from ice-core samples in Greenland and Antarctica. A sharp drop in all three is apparent at the onset of the last ice age. We are presently living in a fairly stable, inter-glacial period which will probably end around 7000 A.D. *(Courtesy of J. Houghton, G. Jenkins and J. Ephraums)*.

the mean surface temperature is only about 10°C lower. Another reason for displaying temperature in this way is to emphasise that we are presently living in an inter-glacial period, and that these warm spells are really quite short in duration, a mere 20,000–30,000 years, compared to the hundred-thousand-years life span of an ice age.

What precipitates an ice age? There are thought to be two principal factors which control the mean surface temperature of the Earth: variations in its orbital path around the Sun and variations in the Sun's energy output. Calculations by astronomers have demonstrated that the Earth's orbit is not stable; rather, it varies in shape from almost circular to elliptical in a cycle lasting approximately 100,000 years (not dissimilar to the duration of an ice age). During this cycle, not only does the Earth's distance from the Sun vary by up to fifteen million kilometres, but our world also precesses (or wobbles) on its axis with a period of about 26,000 years. The present configuration of orbital shape and distance from the Sun favour mild winters and cool summers, but 13,000 years ago it was just the opposite, inducing ice sheets to melt and recede, effectively ending the most recent ice age.

These long-period cycles actually show up in the marine sediment record; isotopic ratios shift in concert with the cycles lasting 26,000 years and 100,000 years. By including all the information known about past temperatures and climate cycles in computer models (chapter 1.3), we can hazard a prediction about the future. If such long extrapolations can be trusted, a new ice age should begin in about 5,000 years.

This may seem contrary to the present concern that human intervention in the climate system may cause the environment to warm up. Might it not be advantageous to let the climate warm over the centuries and millennia ahead, thereby counteracting the advent of another ice age? Well, it would be a long wait, and the climate is likely to change significantly long before that distant time. Consider too the issues raised in Part One, that a change in the energy balance of the Earth-Sun system may alter the circulation patterns of the atmosphere, inferred from the well-established relationship between temperature and wind speeds. Increasing wind speeds, which effectively control the number of storms and the amount and distribution of precipitation, could lead to more extremes in weather. How far can atmospheric flow be disturbed before it acts detrimentally on the climate, shifting it into a new (and perhaps less favourable) configuration?

The effect of changing the mean surface temperature by a mere 5°C would have a devastating impact on the world's harvests, already sensitive to interannual and decadal fluctuations. What would happen if the rain belts moved poleward (some researchers believe this has already started)? If the North American wheat plains were left without adequate water, widespread crop failure would impact strongly on a world where the human population, of which more than two-thirds is already under-nourished, is growing at an almost exponential rate. Higher temperatures and aridity in mid-latitudes might also encourage insect and plant life, and diseases, once confined to the tropics to migrate to higher latitudes.

Do we really have any evidence that we are affecting the heat retention of the atmosphere, and do all scientists agree on the interpretation of this evidence? Our primary source of information about climate comes from the proxy record of past atmospheric composition, extracted from the world's ice sheets. Looking back at the way in which carbon dioxide and methane have varied over the last one-and-a-bit ice ages, both gases have responded in a similar manner, suggesting that their behaviour is closely coupled. When temperatures fall at the onset of an ice age, CO_2 and CH_4 are both locked up in reservoirs, stored by forests on the land and somehow (although no one is exactly sure how this works) by the world's oceans. During these cold periods, CO_2 decreased from around 280 ppmv to around 180 ppmv, and CH_4 from 600 ppmv to 400 ppbv. As temperatures rise again, CO_2 and CH_4 get released back into the atmosphere in a gaseous form, helping to promote the greenhouse effect which warms the planet. Notice, however, that the gases are not released first to end the ice age, since the temperature rises before they are released. The end of an ice age is almost certainly triggered by a change in the Earth's orbit. This is equally true when a cool period begins: as the Earth entered the last ice age 130,000 years ago, CO_2 levels did not start to fall until the air was 7°C cooler; there was a *lag* of several thousand years between a change in temperature and in carbon dioxide.

There is still another reason why we are certain that the near-coincident changes in temperature and CO_2 are not simply cause and effect. If a rise in CO_2 concentrations from 180 to 280 parts per million were solely responsible for inducing a rise in mean surface temperature of 10°C, which is how much the Earth warmed at the end of the last ice age, then the rise in CO_2 from 280 ppmv to 361 ppmv since around 1850 A.D. should have increased the mean surface temperature from 14°C to something like 23°C, whereas it has only risen by around 0.5°C.

Apart from changes in the Earth's orbit, there seem to be variations on shorter time-scales which are induced by fluctuations in the energy output of the Sun. These variations can also be deduced from the proxy temperature record, and thanks to some early astronomers counting sunspots back in the mid seventeenth century, we also have direct visual confirmation of a rather prolonged dip in the Sun's activity. The surface of the Sun is usually peppered with regions which are cool relative to their surroundings. The temperature in these cool regions is typically 4000°C (so they are actually very hot), but because they are 2000°C cooler than the remainder of the Sun's surface they show up as dark patches, which those early astronomers decided to call sunspots.

Sunspots are produced by changes in the Sun's magnetic field which affects the amount of radiation that it emits. In the year 1650 A.D., the surface of the Sun became suddenly quiescent as almost all sunspot activity ceased. This period of quiescence lasted for the remainder of the seventeenth century and several decades into the eighteenth; not until around 1720 did the sunspots reappear in their usual numbers.

There is a well-known solar cycle based on the number of sunspots, an eleven-year cycle (called the *Schwabe* cycle), during which the number of sunspots passes from a minimum to a maximum and back again. However, even during the minimum of this 11-year cycle, there are still a few sunspots present, but during the period 1650–1720 A.D., there were virtually none. Figure 21 shows the annual-mean number of sunspots since the year 1600 (the peaks and troughs reflect the 11-year cycle), together with a thicker curve which is a 10-year running mean value, a technique which largely removes the short-term cycle and accentuates the longer-term changes.

The period between 1650 and 1720, when sunspot activity almost ceased completed, is known as the *Maunder Minimum*. The occurrence of the Maunder Minimum coincides with a cold episode in our climate, usually referred to as the Little Ice Age which lasted from around 1450 until 1800 A.D., a period when summers were cool and wet and winters were unusually cold. Indeed, during several winters towards the end of the seventeenth, and early in the eighteenth centuries, within the span of the Maunder Minimum, the River Thames froze over, the ice sufficiently thick for people to skate on its surface. Agriculture was hit hard in certain

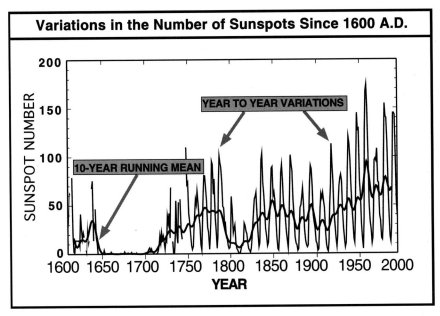

Figure 21 Sunspot numbers since 1600 A.D. The spikes show the annual means, the thick curve a ten-year running mean. The near-total absence of sunspots throughout the Maunder Minimum (1650–1720) is very striking. Sunspot numbers have been steadily increasing throughout the twentieth century, indicative of increased solar activity. Despite this, the solar constant has scarcely changed, suggesting that the rise in activity on the Sun has made only a very small contribution to global warming. *(Courtesy of George Reid, NOAA Aeronomy Laboratory).*

areas of the British Isles, and numerous settlement in Scotland and northern England were abandoned because of repeated crop failure.

The proxy mean surface temperature record for the last four centuries clearly reflects these smaller variations (figure 22A). The years spanned by the Maunder Minimum, between 1650 and 1720, show up clearly. Here, as with the prehistoric record, temperature is shown relative to the present-day surface mean value of 14°C. Notice that the drop in temperature which triggered almost a century of icy winters and cool, wet summers was remarkably small, a mere 0.3°C, less than the amount by which we have raised the mean surface temperature of the Earth over the past century.

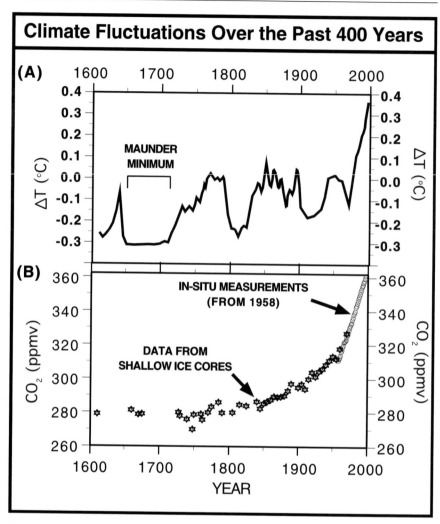

Figure 22 **A**, Deviations from the mean surface temperature, circa 1970, from 1610 A.D. onwards. Notice the cool period corresponding to the Maunder Minimum and later cool episodes at intervals of approximately a century. The increase over the past 30 years is greater than the variation during recent millennia; **B**, changes in the concentrations of atmospheric carbon dioxide, derived from shallow ice-core samples until the 1950s when direct atmospheric measurements began in Hawaii. CO_2 levels are now higher than at any time in the past 160,000 years.

This graph contains another interesting feature: the Maunder Minimum, whilst coincident with the most severe drop in temperature in recent times, is hardly unique; a cool spell tends to occur every hundred years or so. Looking back at the sunspot numbers in figure 21, it is apparent that these cool decades coincide with decreases in the number of sunspots; they are apparently tied to decreases in solar activity. Such a cooling occurred early in the nineteenth century and again, somewhat less distinctly, early in the twentieth century. Interestingly, each successive cool period is a little warmer than its predecessor. Perhaps the rapid warming currently in progress will offset a cool spell early in the twenty-first century.

Until the advent of satellites, it was impossible to detect changes in the Sun's output because of the variable absorption of solar radiation by the atmosphere. Since 1979, however, observations from Earth-orbiting satellites have revealed changes in solar activity on all time scales, from minutes to decades, including a variability associated with the 11-cycle of sunspot numbers. Such changes affect both the amount of radiation reaching the Earth and the distribution of the shorter wavelengths, particularly in the ultraviolet (UV) and X-ray regions of the spectrum.

The variation in atmospheric carbon dioxide concentrations during glacial and inter-glacial periods have remained between 180 and 280 ppmv, a range of 80 ppmv. In less than two centuries, CO_2 has increased by this much again, reaching 361 ppmv in 1997 (figure 22), a value far in excess of anything recorded for the past one hundred and sixty millennia. The increase is continuing at a rate of between 1 and 2 ppmv each year, and this is expected to accelerate in the future (almost exponentially unless emissions are curbed). By the early 1990s, between 6 and 7 gigatonnes (6–7 billion tonnes) of carbon dioxide were being added to the atmosphere each year from the combustion of fossil fuel. Not all the emitted CO_2 remains there, of course; roughly half is absorbed in natural sinks such as growing forests and the Earth's oceans.

Returning once more to the variations in sunspot activity over the last 400 years (figure 21), there is, in addition to the 11-year and 100-year cycles, a steady increase in sunspot abundance throughout the twentieth century. The Sun is apparently more active now than it was a few centuries ago, and just as the Maunder Minimum resulted in a drop in global temperatures, this enhanced activity may very well induce an increase. The present production of the carbon-14 (C^{14}) and beryllium-10 (Be^{10}) appears to be near historically-low levels because of persistently high solar activity, which inhibits the rate of production of these isotopes. A dip in the proxy record of both C^{14} and B^{10} during the 12th century indicates when solar activity was last at a maximum, a time called the *Medieval Maximum*. As yet, we cannot predict with any certainty when such maxima (or minima) in solar activity will occur, not even within the next 11-year Schwabe cycle, which makes predicting the climate impact of future changes in solar output virtually impossible. That a change in solar output would affect the climate seems certain: when an 11-year mean of the annual-average northern hemisphere surface temperature is compared with variations in the length of the Schwabe cycle, a strong correlation is found. Indeed, since 1850, variations in the mean surface temperature of the Earth seem to have moved in concert with variations in solar activity equally as well as they have followed increases in greenhouse gas concentrations. According to current opinion, however, changes in solar activity can only account for around half of the warming which has occurred since that time, and for only 30% of the warming over the past 30 years. At the very least, it complicates the unambiguous detection of a climate change signal induced by human activities.

Let us assume, for the sake of argument, that we *are* making a tangible contribution to the present upward trend in global temperatures. What does this mean for policy makers as they try to plan for the future? There are moves afoot, in the form of the Kyoto Protocol (see chapter 3.4) to make the present 39 *developed* countries limit emissions of key greenhouse gases, but given that our principal form of energy production is still carbon-based, such a move is likely to impact unfavourably on the economies of these countries and meet with considerable resistance. And even if we do ultimately agree to limit emissions of gases such as CO_2 and other selected species, the problem of climate change remains far from solved.

The fastest growth rate of the human population is now in the developing countries; at the present time, and doubtless for many decades to come, these countries can scarcely afford to worry about the esoteric effects of global warming. The danger is that poorer nations are only just beginning to utilise the technologies which have been available to their wealthy neighbours for the past century or more, and because of their spiralling populations they are expected to contribute to the greenhouse effect on a monumental scale. By around the year 2010, we may see third world

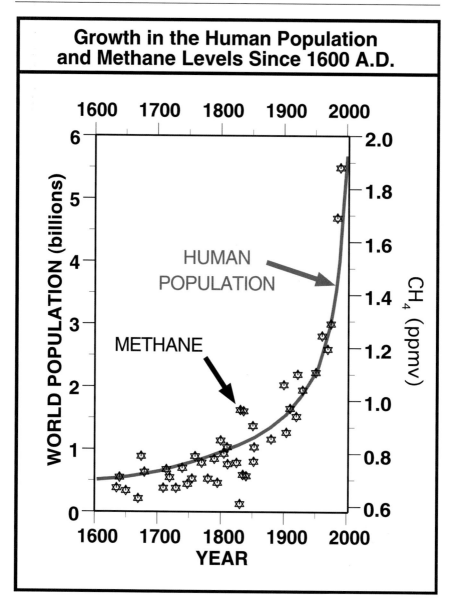

Figure 23 The growth in methane concentrations in the atmosphere from circa 1600 A.D. (shown as stars) mirrors that of carbon dioxide shown in the previous figure. The solid curve indicates how the human population has grown, almost exponentially, during this same period.

greenhouse gas emissions eclipsing those of the developed countries. At the time of writing, these nations are not yet included in measures to limit future emissions, and without their participation, any measures adopted by the developed nations are unlikely to be effective.

Figure 23 emphasises the dilemma: the grey curve shows how our population has climbed over the past four centuries, most of the increase occurring in the twentieth century alone, with almost a doubling since 1970. By the year 2050, our numbers are expected to reach something like 9.5 billion, with more than half of this growth on the Asian continent. Included on the same figure (as stars) are proxy measurements of methane, the second most important greenhouse gas being emitted into the atmosphere, for which (until recently) the developing countries were primarily responsible.

OCEANS AND CLIMATE: EL NIÑO AND LA NIÑA

In the previous chapter, it was explained how the global mean temperature has risen by about 0.5–0.6°C (~1°F) during the past hundred years. Moreover, the last 10 years (1987–1997) have been the warmest on record. The widespread melting of glaciers and thermal expansion of the oceans have contributed, it is believed, to a rise in sea level of something like 15 cm (6 inches) over the past century.

The vast body of water enshrouding more than 70% of the Earth's surface has an enormous thermal inertia, taking a surprisingly long time to respond to change, and an equally long time to return to its former, equilibrium state. This means that changes made today are likely to affect us not so much in a few years but on time-scales of decades to centuries, and if significant changes are set in motion it will be impossible to stop them.

From the discussion of large climate variations associated with ice ages in chapter 3.2, it is evident that the onset of cold periods can be triggered very rapidly: temperature changes of 5–10°C can take place in just a few centuries, sometimes over just a few decades. Generally speaking, the coldest parts of the ice ages and the present period of warmth are comparatively stable, although this was not strictly the case during the most recent ice age. Near the onset and end of an ice age, however, the climate hops back and forth between warmth and cold, a behaviour climatologists call *climate flickering*. During this series of abrupt returns to a glacial climate, the best known of which was the *Younger* Dryas event that lasted from around 11,000 to 10,000 B.C., temperature shifts of 7 to 10°C occurred over just 50 years or less.

In an effort to explain this apparently erratic behaviour, scientists have begun looking beyond variations in the Earth's orbit and solar energy output. It is now believed that a change in climate in one geographic area of the Earth's surface, however induced, may trigger variations in another. For example, a localised warming at high latitudes might very well influence the circulation of both the atmosphere and the oceans in mid-latitudes, which would ultimately lead to a warming in the tropics. If the tropics are warmed, the amount of evaporation from the surface of the oceans will increase, injecting more water vapour into the atmosphere and probably increasing global cloud cover.

Until recently, little consideration was given to the relationship between atmospheric and oceanic circulations, but the two are in fact very strongly interconnected. Even in the early 1980s, our understanding of the rôle played by the oceans in the climate system was mostly descriptive, and the available data were inadequate to test the predictions generated by models designed to simulate this relationship.

In chapter 3.1, it was stated that climate may be viewed as meaning *the average physical environment of the Earth*. The primary contributors to this average physical environment are the atmosphere and oceans. The atmosphere absorbs about 70% of incoming solar energy, whilst the uppermost 20–30 metres of the oceans absorb most of what reaches the surface. This, not unexpectedly, warms the surface waters, especially in the tropics where the Sun is overhead and a thinner atmosphere intercepts less ultra-violet radiation than at higher latitudes. Because sea water possesses a high thermal capacity (an ability to store heat), and because ocean currents drive water over huge distances, these large bodies of water dominate the redistributing of heat around the Earth.

Ocean currents are essentially driven by two mechanisms. In the upper part, close to the surface, currents are created and maintained by the wind; the oceans and the atmosphere carry on a continuous dialogue, each listening to what the other has to say and responding. Winds are responsible for the strong currents which flow poleward from the subtropical basins, located around 10° and 40° latitude in both hemispheres. One of these currents, the North Atlantic Gulf Stream keeps the east coast of North America considerably warmer than it would otherwise be, whilst the North Atlantic Drift does the same for Europe. This poleward flow returns to the equator in slower, broader currents over a number of years. Intimately linked to the wind-driven currents is a slower but climatically more important current called the *Thermohaline Circulation* which cools warm surface waters in the polar regions. This cooling makes the water denser, inducing it to sink to depths of several kilometres before flowing towards the equator. The speed of this circulation, estimated by sampling sea water far below the surface, is remarkably slow; it takes the water between 3 and 4 decades to reach the subtropics. This long-period circulation, carrying warm water poleward and cold water equatorward, has the net effect of transporting heat towards the poles.

There are other currents which have cycles lasting thousands of years, and it is these which are likely to carry the impacts of climate change far into the future. The important question is precisely *how* the warming of the oceans will affect the climate? If the oceans warm, even by a few degrees, two things are expected to happen in concert: we know from everyday experience that if we heat a pan full of water, it will expand to fill a larger volume. Similarly, we can expect the oceans to expand, and since they cannot expand laterally, they will rise. A second contribution would be the release of water from polar ice, which a warmer ocean would melt from below. As already mentioned, the mean sea level around the planet is believed to have risen 15 cm in the last century alone, and that glaciers around the world have been rapidly receding since the mid-nineteenth century.

The rate at which sea level will rise is governed by a number of factors: using the best estimates of climate sensitivity parameters, including the cooling effect of aerosols in the atmosphere and ice melt from the world's glaciers, as well as the thermal expansivity of sea water, three different scenarios have been modelled. Curve A in figure 24 assumes a climate sensitivity of 4.5°C (that is, an increase in mean surface temperature of 4.5°C during the 21st century); curve B a sensitivity of 2.5°C, and curve C, 1.5°C. The worst case, curve A, shows a rise in mean sea level of almost one metre by the year 2100. This may not sound like very much, but the increased *range* between low and high tides, of the order of several metres, would be devastating, especially for countries which already experience problems with sea defences. This projection is for a single century, but at present there is every indication that the effects of global warming will continue far beyond this time, unless counter-measures are taken in the meantime.

The huge natural variability in climate which takes place on long time scales, discussed in the previous chapter, is linked primarily to the Sun's energy output. Superimposed on this are some short-term variations in the ocean-climate system which, for the most part, are still not well quantified. One of these short-term variations has received a great deal of publicity in recent times,

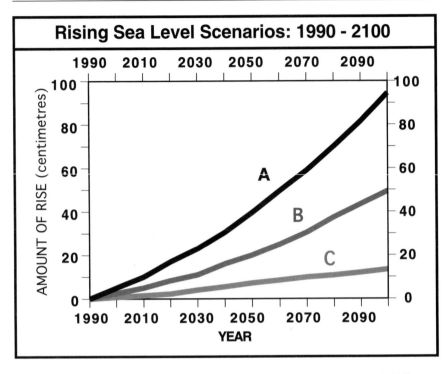

Figure 24 Rising sea level scenarios based on predictions of different levels of global warming. An increase of almost 1 metre (relative to 1990 levels) is expected during the twenty-first century (Curve A) if the mean surface temperature rises by 4.5°C during this period. The scenarios vary according to how much greenhouse gases are expected to increase, and the extent to which the polar ice sheets melt (see text).

an event which occurs every few years, during which regions of the tropical Pacific warm significantly and affect the climate over much of the planet. It is, of course, El Niño.

The name El Niño, which may be translated as *The Little Boy*, or *The Christ Child* (because of its tendency to appear around Christmas time) holds a romantic appeal, but it conveys no information about the true nature of the phenomenon. El Niño (abbreviated **EN**) is a periodic shift in ocean temperatures in the tropical Pacific Ocean, normally accompanied by a change in the atmospheric circulation known as the Southern Oscillation (discussed in connection with future climate in chapter 3.4). The southern oscillation (abbreviated **SO**) is called so because it oscillates between a warm and cold phase which is linked to the comings and goings of El Niño itself, and because of this interrelation scientists generally refer to the combined event as an El Niño-Southern Oscillation, or ENSO.

At first glance, this warming of the ocean and shift in atmospheric circulation seems fairly localised, confined to the southern Pacific. Indeed, for many years, it was largely ignored because of its apparently localised impact. More recently, we have come to appreciate that its presence is felt in places far removed from its source region.

Table 6 El Niño and La Niña years in the twentieth century.

(a) El Niño Years			
1902–03	1905–06	1911–12	1914–15
1918–19	1923–24	1925–26	1930–31
1932–33	1939–40	1941–42	1951–52
1953–54	1957–58	1965–66	1969–70
1972–73	1976–77	1982–83	1986–87
1991–92	1994–95	1997–98	
(b) La Niña Years			
1904–05	1909–10	1910–11	1915–16
1917–18	1924–25	1928–29	1938–39
1950–51	1955–56	1956–57	1964–65
1970–71	1971–72	1973–74	1975–76
1988–89	1995–96		

An El Niño warming takes place every 4.9 years on *average*, but with a variability of 2 to 7 years (see Table 6 for a list of El Niño years this century). Whilst we can predict it with moderate certainty, its actual onset and severity still defeat us. The El Niño which occurred during the winter of 1982–83, one of the most powerful warmings of the twentieth century, left thousands of people dead and caused $13 billion (U.S.) (approximately eight billion pounds) worth of damage to property in the United States alone. Predictions for the 1997–98 El Niño indicated this would probably be even more devastating, although in the early part of the winter virtually nothing happened, emphasising the difficulty with forecasts of changes just a year in the future.

As events transpired the anticipated havoc came to pass, albeit a little later than expected. The number of hurricanes in the central Pacific escalated, and in February 1998 California felt the presence of Hurricane Linda as it traversed the eastern Pacific, bringing with it winds in excess of 300 kilometres per hour, the strongest ever recorded in this region. Ten-metre-high waves and torrential rain swept landward, destroying thousands of houses and bringing public services to a standstill. In fact, California was hit by not one but *three* major storms in the space of a single week. Far to the east on the Atlantic seaboard, tornadoes cuts swathes through parts of Florida, whilst in the interior of the continent, some areas of the United States experienced the warmest winter in decades.

The effects of El Niño were also felt in parts of South America which experienced flash floods, whilst in Indonesia and Australia forest fires raged out of control for months as the worst drought in half a century took hold, and more than a million acres of Indonesian forest were lost. In stark contrast, east Africa, victim of so many droughts in recent decades, was subject to torrential rainfall and extensive flooding.

What is it that triggers this warming of the upper layers of the tropical Pacific? Once again, the description of the general behaviour of the atmosphere in Part One will prove useful. In *normal*, non-El Niño years, the Trade Winds blow from east-to-west across the tropical Pacific, driving the sea water before them and piling up warm water in the western Pacific (**Plate 3a**). Consequently, the sea surface is about half a metre higher and up to 10°C warmer near Indonesia (in the west) than near Ecuador (in the east). The east is cooled by an upwelling of cold water from deeper levels in the ocean which are nutrient-rich, supporting a diverse marine ecosystem and the fishing industry.

During an El Niño event, however, the Trade Winds slacken in the central and western Pacific, it is believed because of high pressure on the eastern side. This leads to a depression of the thermocline in the east (that is, the temperature gradient of the ocean over an arbitrary distance decreases) and an elevation (steeper gradient) in the west. The upwelling of deep waters, which normally cool the surface in the east, becomes less pronounced with the result that the sea surface temperature near Ecuador *increases* by as much as 10°C (**Plate 3b**) and the water becomes nutrient poor, affecting organisms further up the food chain – fish populations decrease because there is now less food.

Rainfall follows the warm water eastward, typically with associated flooding in Peru and drought in Indonesia and Australia.

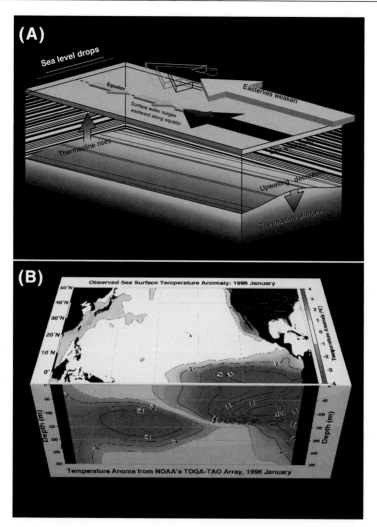

Plate 3　(a) El Niño is triggered by a slackening of the easterly (from the east) winds in the equatorial Pacific, decreasing upwelling and increasing the surface level on the eastern side of the ocean, and correspondingly increasing upwelling and decreasing the surface level on the western side; (b) the weakened easterlies (winds from the east) cool the eastern ocean surface less and allow the waters to warm up, whilst in the west the waters tend to cool. This creates a strong temperature gradient allowing the surface waters to plunge westward. This movement alters the path of the subtropical jet stream, and this change in the weather pattern is communicated to other regions of the atmosphere. *(Courtesy of D. Pierce, Scripps Institution of Oceanography).*

Why? The eastward displacement of the atmospheric heat source overlaying the warmest water generates large changes in the global atmospheric circulation which in turn force weather changes in regions far removed from the tropical Pacific. The modification of the atmospheric circulation arises essentially because the sub-tropical and mid-latitude (polar front) jet streams are displaced from their usual paths. In the northern hemisphere, for example, the polar jet stream is displaced poleward so that the cold and dry Arctic air does not reach the eastern United States, while in the Pacific, the subtropical jet stream moves equatorward and becomes more intense. It was this displacement which was responsible for ducting the series of major storms towards the Californian coast early in 1998. In mid-latitudes, low pressure systems tend to become more vigorous than normal, especially in the vicinity of the Gulf of Alaska. These weather systems pump abnormally warm air into Alaska, western Canada and the north-west United States. Storm strengths are similarly enhanced in the Gulf of Mexico and along the south-eastern coast of the United States. These regions are the locations of the strongest storm tracks (regions where strong *diabatic* heating occurs) on the Earth.

What is diabatic heating? In Part One, we learned about quantities which do not change rapidly with time (potential temperature and potential vorticity) which permit us to follow the path of an air parcel through the atmosphere. These invariants rely on the assumption that the air does not exchange energy with its surroundings; this is called the *adiabatic approximation*. When the behaviour of the air is *diabatic*, energy is exchanged with its surroundings and strong vertical motions occur, releasing large quantities of heat and promoting the development of storms.

Whether El Niño is present or not, storm tracks do not form without additional influences, and as with most atmospheric disturbances large-scale atmospheric waves are the underlying mechanism. Such waves perturb the flow of the atmosphere very effectively over the oceans, inducing the strongest storm tracks in mid-latitudes. There is one such track in the mid-latitude Atlantic which causes the high frequency of storms which assail Florida, whilst another (generally much weaker) one exists over the central Pacific. Because of the interconnecting nature of oceans and storms, warming the oceans is likely to affect the intensity, and perhaps even the positions, of these storm tracks in the future.

Returning to El Niño, why does this warming event occur only in the Pacific and not in the Atlantic or Indian oceans? Most of the current theories which try to explain the mechanisms under-pinning El Niño warmings involve planetary-scale waves crossing the Pacific. The key seems to reside in the time taken for these waves to cross the Pacific, which apparently plays an important rôle in setting the time-scale and amplitude of the warming anomaly. The narrower widths of the other oceans seems to pre-vent the same thing happening there: because the planetary waves cross the Atlantic and Indian basins in less time, these oceans can adjust more rapidly to variations in the wind. In addition, the bordering land masses heat and cool in response to the annual migration of the Sun and tend to overwhelm the oscillatory be-haviour which triggers an ENSO event. In the Pacific, however, where the adjustment time is slower, the ocean-atmosphere system drifts further from equilibrium and the adjustment back to equi-librium conditions takes a number of months.

In between El Niño warmings, which typically occur every 2–7 years, there is sometimes (but not always) a cooling event called La Niña, Spanish for *The Little Girl,* (an alternative name sometimes used is *El Viejo*). These years are characterised by unusually cold ocean temperatures in the equatorial eastern Pacific, when a cool tongue of water extends farther west that usual. The effect on climate is to promote a temporary cooling over some regions of the Earth.

As suggested already, El Niño and La Niña tend to alternate, the interval between each event varying significantly. Given their relatively short life-times, coupled with our inability to forecast them accurately, it is hardly surprising that computer models have difficulty in predicting the future climate of the Earth, say fifty years ahead. The models themselves are, after all, full of uncer-tainties (Chapter 1.3).

Evidently, a comparatively small change in the ocean-atmos-phere system is required to set up an El Niño. This suggests that even fairly small perturbations to this system will affect the strength of the circulations, disrupt established weather patterns and alter, for example, the timing of the monsoons and other precipitation events. Proxy climate records reveal that El Niño does vary in response to temperature-circulation changes, and we know that El Niño is a phenomenon which has been operating for at least the past two million years, although not continuously. When the last ice age was drawing to a close (between 14,000 and 9,000 years ago), ENSO apparently did not occur at all. This was a time when ocean levels were rising dramatically, and this may have served to reduce atmospheric pressure differences, preventing El Niño from getting started.

The purpose here in focusing on ENSO in such detail is to highlight the way in which a comparatively small change in the strength of the winds blowing across the Pacific can induce substantial changes in ocean currents, which in turn impact on weather world-wide. If the planet is warmed in response to either natural or anthropogenic influences, extremes in weather are very likely to become more severe.

At the National Center for Atmospheric Research (NCAR) in Boulder, Colorado, scientists have recently constructed the first climate model which attempts to incorporate the El Niño Southern Oscillation event, hoping to predict changes which might result from an atmosphere containing twice as much carbon dioxide as it did around 1800 A.D. The model predicts that weather anomalies connected with ENSO would indeed become more extreme, with the amount and distribution patterns of precipitation changing dramatically.

If we are compelled to live in a world where the oceans are warmer, in which we may expect more extremes in local climates, then we must also face the fact that there may be a shift of power in the world's economies. For example, if the number of storms increase and rain belts continue to shift polewards, the North American wheat plains and much of the African continent may be wracked by extreme weather conditions which, year after year, could destroy crops by deluge or drought. Even Britain, renowned for its rainy climate, has experienced less rainfall on average in the past several decades, leaving some areas short of water and experiencing drought conditions, especially in the south-east.

Can we be certain that changes in climate are on the way? If we take preventive action, might we not be squandering huge sums of money trying to pre-empt them, and after all our efforts still fail? Even if changes are on the way, there is still the question of whether or not they have arisen from human activities. Perhaps, after all, they are merely a part of the climate's huge range of natural variability. These are questions which have attracted much debate and controversy. We know there is a natural variability in both the Earth's orbit and solar activity, as well as in other cycles which work on time-scales ranging from decades to centuries. So is there really any unambiguous evidence that mankind is making a significant difference?

There are those who have serious reservations about the accuracy of the land-based temperature record, for which direct measurements go back only as far as 1850 A.D. Satellite records, which go back a mere two decades, are equally uncertain. These have recently been revised by Christie et alia (July 1998), but even this has been challenged by a further scientific paper (still unpublished at the time of writing). The latest controversy surrounds satellite orbital decay corrections which are described by Wentz and Schabel (Nature, 1998): previously-used corrections, it has been argued, are responsible for a spurious cooling of the order of ±0.1°C per decade since 1979, and that new corrections actually produce a warming trend. This serves to underscore an important point: the magnitude of the corrections to the satellite data record have a range of uncertainty of ±0.1°C, far larger than many published claims that trends can be determined to an accuracy of ±0.05°C.

There is another important factor to keep in mind when comparing the ground-based and satellite temperature records. Ground-based measurements are nearly all made on land, whereas satellites measure temperatures over the *entire globe*, much of which is covered by water. Ground-based measurements show a fairly clear warming trend, most of which arises from increased night-time temperatures. When satellite data over the land masses only are examined, they too show a warming in progress. But then, over the oceans, they detect a cooling which is at odds with the estimated rise in sea level during the twentieth century. Furthermore, this does not accord with observations of retreating ice around the world: most ice loss occurs via the calving of icebergs, but a significant amount at the base of ice shelves is melting, a fact discovered by drilling down through the ice sheets and comparing, over a period of time, the depth at which the ice becomes sea water.

Should we trust satellite data at present? Probably not too much, and certainly not before the various corrections have been calculated to everyone's satisfaction.

POSSIBLE FUTURES

In this chapter, as with the ozone issue in chapter 2.4, we shall examine the principal factors which either will, or *may*, contribute to anthropogenic climate change, and then consider what we can do to minimise their effects. The Kyoto Protocol, drawn up in Japan in December 1997, is the first legal document to propose regulating the emissions of greenhouse gases into the atmosphere, but it has a number of shortcomings, not least among them the exclusion of the influence of aircraft emissions on climate. A major drawback at this stage is that we still do not have all the information we need to make sound decisions about the future. For instance, the Kyoto Protocol calls for a limitation on the growth rate in carbon dioxide emissions, but in order to appreciate the effect of elevated CO_2 concentrations on future climate, we need to properly understand the carbon cycle, something we are still far from achieving.

(i)   The Case of the Missing Carbon

Every year, we release approximately seven billion tonnes of carbon dioxide into the atmosphere. Since the atmosphere presently contains around 730 billion tonnes, our contribution seems rather trivial. But there is a mystery: at the present time, the burning of fossil fuels and the manufacture of cement account for 5.3 billion tonnes of CO_2 emissions, or *sources*, and deforestation for between 1 and 2 billion tonnes, summing to 6.3–7.3 billion tonnes. We know that the atmosphere stores some 3.3 billion tonnes of this annually, whilst the oceans soak up a further 2.1 billion tonnes. These are known as CO_2 sinks, mechanisms which *remove* carbon dioxide from the atmosphere. We are unable to account for the difference of between 0.9 and 1.9 billion tonnes each year, however. Because of our present lack of understanding, we face two potential dangers: there may come a point when the oceans warm sufficiently to inhibit their ability to store CO_2, and there may come a point when the missing CO_2 sink is no longer effective in a warmer world. In either event, some or all of the unaccounted annual difference of 1–2 billion tonnes may be left to build up in the atmosphere.

Such changes, if they occur, will probably be in response to temperature changes associated with ice ages. In chapter 3.2, it was demonstrated how both carbon dioxide and methane are sequestered during cold periods in the Earth's history; at these times, the atmospheric concentration of CO_2 falls from around 280 to 180 parts per million, and CH_4 from 600 to 400 ppmv. When the atmosphere warms again at the end of an ice age, CO_2 and CH_4 are released back into the atmosphere in gaseous form. This positive feedback mechanism promotes a warmer atmosphere which then liberates more carbon from the oceans, which in turn warms the atmosphere still further and soon.

Let us consider what the future may hold when we have finished removing the world's rain forests. Most carbon is taken up by young growing forests rather than the old (boreal) forests, which on the face of it implies that removing the rain forests would have little effect on the carbon budget. However, when rain forests are removed, they are not replaced with new growth; the cleared areas are used for short-term agriculture until the fertile soil, no longer locked in place by plant life, is washed away by the rains. Without trees, the amount of precipitation rapidly decreases. The capacity of trees to take up carbon is strongly influenced by the amount of precipitation, with lower rainfall leading to lower carbon uptake. Moreover, if, or when, the climate warms, the timing and distribution of precipitation is likely to change. The consequences of removing the rain forests may be far reaching indeed.

Other factors, such as the amount of cloud cover, snow depth and the length of the growing season also impact on carbon uptake. Recent research by Dr. Steven Wofsy at Harvard University suggests that whilst warmer weather increases the CO_2 uptake in young forests, it promotes its release from older forests, and that the warmer years over the past few decades have triggered the release of old CO_2 which had long been locked up in soil. Perhaps the greatest concern is that carbon reservoirs can be quickly mobilised by temperature change. For example, if soil does not stay water-logged, carbon is released, so that areas of the planet which have suffered persistent droughts, or which may suffer them in the future, have the potential to release substantial amount of carbon into the atmosphere.

(ii)   The Impact of Subsonic Aircraft on Climate

Since the early 1970s, considerable attention has been directed towards the ozone depletion potential of supersonic stratospheric aircraft, but comparatively little attention has been devoted to influence of the present (and future) subsonic fleets. According to a recent report by the United Nations Inter-Governmental Panel on Climate Change, subsonic air traffic consume about 3% of the total fossil fuels world-wide, with an expected increase up to 10% by the middle of the twenty-first century. The report goes on to state that aircraft may already be responsible for some 5–6% of the global warming caused by greenhouse gases. Aircraft emissions, it claims, are doubling every ten years or so, and account for more than half the global warming potential (Appendix 3.1) of emissions from ground transportation.

Aircraft also generate active nitrogen species — nitrogen monoxide (NO) and nitrogen dioxide (NO_2), collectively called NO_x; the oxides of nitrogen, as well as nitrous oxide (N_2O). Various studies, summarised in the World Meteorological Organisation's ozone assessments (see bibliography for Part Two) indicate that aircraft release most of the NO_x species at their cruise altitudes between 9 and 13 km; into the upper troposphere. A recent report by European scientists suggests that aircraft are responsible for 1–2% of the total nitrogen oxide concentrations observed.

In the upper troposphere, NO_x serves to create ozone rather than destroy it (as it does in the lower stratosphere). Because it too is a greenhouse gas, ozone at these altitudes traps radiation and can potentially contribute to global warming. However, whilst ozone is unquestionably increasing in the upper troposphere, the precise contribution of subsonic aircraft has yet to be established to everyone's satisfaction. Paul Wennberg at the California Institute of Technology in Pasadena, a co-author of the latest IPCC, reports in its first chapter that ozone is probably being produced at twice the rate assumed in the report, based on work he did on data collected by NASA's ER-2 research aircraft during 1995 and 1996. Apparently, the aircraft sensors recorded far more hydroxyl (OH) than expected in the upper troposphere, and it is this radical which oxidises NO_x species to ozone via photolysis. These high hydroxyl concentrations cast doubt on the long held assumption (currently used in climate models) that OH derives mainly from water. Since the upper troposphere is very dry, the high levels of

OH suggest that a suite of chemical reactions is taking place which involves substances other than water.

The regions of the atmosphere most heavily-used by subsonic aircraft, as of 1992, are the European and North American continents, and the North Atlantic Flight Corridor. This is indicated in plate 4a, which shows NO_x emissions (integrated between the ground and 16 km) during July and November 1992, when there were 595 and 462 trans-Atlantic flights (in both directions), between 45°N and 60°N. The scale on the colour table on the right of the plot is an emissions index (**EI**) which relates the amount of nitrogen dioxide emitted per kilogram of fuel burnt. This information for the year 1992 is summarised in Table 7. **Plates 4b** and **c** show modelled NO_x emissions for the years 2015 and 2050, respectively. Notice the shifting colour scale to the right of the graphs: the red band in **Plate a** (1992) represents an emission index in the range 1–5, in **b** (for the year 2015) 7–9, and in **c** (for the year 2050), 20–100.

Table 7   Annual fuel burnt and NO_x emitted (as nitrogen dioxide) from all aviation sources, forecast for the year 1992.

Categories of aircraft	Fuel burnt (in tg* per year)	Fuel burnt (in tg* per year)	Emissions Index for NO_x
Commercial subsonic	272.3	272.3	12.4
Military	14.5	14.5	10.7
Total civil and military	286.8	286.8	12.3

* tg = a teragramme = one-thousand-billion grammes.

There are a number of ways in which aircraft are believed to affect the radiative forcing (Appendix 3.2) of the atmosphere. For instance, aircraft contrails are a familiar sight in skies around the world. As they disperse, they may easily be mistaken for clouds, and in many respects that is exactly how they behave. Models indicate that a constant cover of optically-thin contrails causes a warming at the top of the troposphere and a cooling at the Earth's surface. This effect makes a contribution to the radiative forcing of the climate of around 0.02 watts per square metre

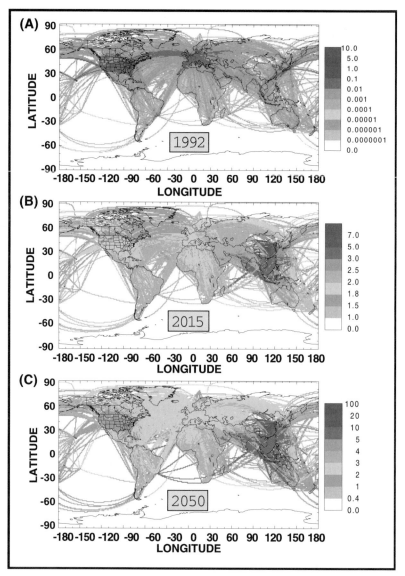

Plate 4 Spatial distribution of 1992 NO$_x$ emissions (A) from civil aviation, vertically integrated between the ground and an altitude of 16 km, in units of kilograms of NO$_2$ per metre-squared per year (kg NO$_2$ m^{-2} yr^{-1}); modelled spatial NO$_x$ distributions for the years 2015 (B) and 2050 (C), compared to 1992. Note the changing colour scale to the right of each figure, on going from A to C. *(Courtesy David S. Lee, A.E.A. Technology plc, Oxfordshire).*

(Wm^{-2}). Some aircraft manufactures have been working to reduce emissions by improving combustion, cutting weight and refining the aerodynamics of their aircraft. Engines with lower emissions will lead to lower operating costs, and so are beneficial both to the environment and to the industry itself. The American company General Electric has been collaborating with the military and NASA to develop a Duel-Stage Combustor engine which will only generate a small fraction of the hydrocarbons, carbon monoxide and NO$_x$ emissions currently permitted by the International Civil Aviation Organisation (ICAO).

Recent research (by Brian Toon of the University of Colorado in the U.S.A. and many others) has drawn attention to the difficulty in distinguishing between an aircraft contrail and a deck of cirrus cloud. The ambiguity arises because, after a few hours, contrails spread out and become indistinguishable from clouds, at least to the naked eye. It turns out that contrails may impact sufficiently on cirrus to become radiatively significant, particularly in the future as the number of aircraft increases. Research in Europe (e.g., Sausen et al., 1998) suggests that the amount of cirrus may already have increased by 0.1% over the entire northern hemisphere, and by up to 5% in localised areas. They estimate that by 2050, around 10% of Europe and the U.S.A. may be covered by contrails, and hence cirrus cloud.

Increasing the amount of cirrus would alter the albedo of the Earth. Indeed, the Earth's albedo is believed to have changed significantly over the past billion years or so, and some of the change being recorded now may be caused by a shift in climate. If the particle sizes in clouds become smaller, the albedo will increase and the Earth will cool, but if they grow larger (they can become sufficiently large that no clouds would be visible at all), then the Earth will warm.

The relative amount of climate forcing by different mechanisms is summarised in Appendix 3.2.

Two major aircraft assessments on the impact of aircraft on climate are in preparation at the time of writing: the 1998 NASA Assessment of High Speed Civil Transport, and the 1999 IPCC Special Report on Aviation and the Global Atmosphere.

(iii) Natural Changes in Atmospheric Circulation

In addition to providing insights into the climate of the distant past, the Greenland ice-core data have also revealed the existence of large, short-term climate variations which typically last for one or two decades (they are actually known as *decadal* variations). The mechanisms which control these variations remain poorly understood, but what is clear is that they are linked to the winds blowing across the surfaces of the oceans, especially in the North Atlantic and equatorial Pacific. These winds induce differences in sea level pressure (SLP) over distances of thousands of kilometres. The situation is further complicated by the fact that such decade-long variations are often referred to as oscillations, because of their cyclic nature. The Atlantic and Pacific cycles are known as the *North Atlantic Oscillation* (NAO) and *Southern Oscillation* (SO), respectively. The SO was mentioned in connection with El Niño (EN) in chapter 3.3, and because El Niño plays such a major rôle in controlling the SO, the two are generally linked together and referred to as ENSO.

By comparing recent mean surface temperatures and mean surface pressures with earlier years, it is possible to see how some parts of the globe are anomalously warm whilst other parts have cooled. In **plate 5a**, the mean surface temperatures for the period 1981 to 1996 have been subtracted from the mean surface temperature for the period 1951–1980, with the result that Europe, Russia and North America all show up as being warmer than they were a few decades ago (the red areas on the graphs). Conversely, the southern Pacific and the North Atlantic (between Greenland and Europe) oceans are cooler now than earlier this century.

Turning to sea level pressure (**plate 5b**) as an indicator, it is apparent that the Atlantic ocean surface pressure has been lower in recent years than previously, whilst in the polar and mid-Pacific oceans it is higher. We shall return to **plate 5c** shortly; first, let's consider how the magnitude of these oscillations in atmospheric flow are measured.

Scientists have constructed yard-sticks to gauge the strengths of the SO and NAO which they call *indices*. All this mean is that differences in temperature, or sea level pressure (SLP), are obtained by subtracting data collected at one location from data collected at another, the sites themselves being separated by thousands of kilometres. To get the southern oscillation index (SOI), SLP data from Darwin (12.4°S, 130.9°E) in northern Australia, are subtracted from data at Tahiti (17.5°S, 149.6°W). The result is shown in **plate 5d**. The same technique has been used for the North Atlantic Oscillation Index (NAOI), which acts more

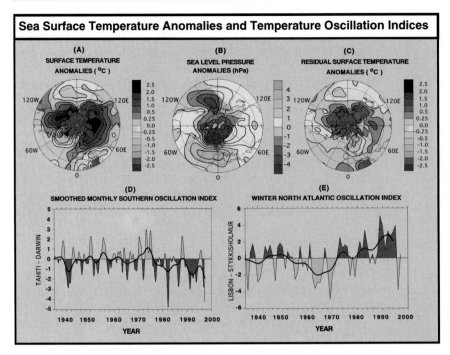

Plate 5 Persistent anomalies in surface temperature (A) and sea level pressure (B) caused by shifts in atmospheric circulation. The residual temperature anomaly (C) cannot be explained by this natural mechanism. Persistent changes are gauged by comparing temperatures at widely-separate sites in the southern (D) and northern (E) hemispheres, a method to which scientists refer as an index (it is actually a pressure index). The red regions in (D) and (E) suggest a persistent warming (compared to earlier years) which has been in progress throughout the last quarter of the 20th century. (Courtesy of James Hurrell, National Center for Atmospheric Research).

in the north-south direction: here, SLP data at Stykkisholmur in Iceland (65.1°N, 22.7°W) are subtracted from those at Lisbon, Portugal (38.8°N, 9.1°W) **(plate 5e)**.

What do these indices tell us about the circulation of the atmosphere? When ENSO is in a warm phase (the red areas in **plate 5d**), the central and eastern tropical Pacific water shifts the location of the heaviest tropical rainfall from Indonesia to the far western Pacific. The warm phase, then, corresponds to an El Niño warming event which, as demonstrated in chapter 3.3, profoundly affects weather patterns across much of the globe.

Whilst the southern oscillation has a period of 2–7 years, its strength varies considerably from one event to the next; that is, it exhibits an *inter-annual variability*. Changes lasting for a decade or longer (decadal changes) show up as a persistence of one phase (either warm or cold, corresponding to red or blue in the figure). Between 1930 and 1950, the year-to-year variations were quite small, except for the event during 1939–42. What is remarkable is that the decadal (and longer) variations of recent years are entirely lacking from earlier periods. Particularly noteworthy is the shift towards more negative values (and thus warmer conditions in the tropical Pacific) since the mid-1970s, shown by the preponderance of red areas in **plate 5**. After 1990, apart from a very short-lived cool period in 1996, the index has been continuously negative, which is unprecedented in the 140-year-long record of these measurements.

A similar story has emerged from the NAOI, where positive values indicate stronger-than-average westerly (from the west) winds over mid-latitudes, corresponding to warmer conditions. This is still represented by the red areas on the graphs, but they have switched sign. Despite the large inter-annual variability, as with the SO, there have been several periods when the NAOI has persisted in one phase over many winters. For example, from the 1950s until the early 1970s, temperatures over Europe were lower than average (corresponding to the negative, or blue, regions), whilst from 1980 onwards, temperatures have been higher (corresponding to positive, or red, regions).

These persistent phases (positive in the NAOI and negative in the SOI) have a lot to do with why land temperatures have been high during the past few decades, and this has unquestionably contributed to the trend in the mean global temperature. The cool period around the middle of the century (especially in the NAOI), and the warm period towards the end of the century can account for some (but certainly not all) of the global warming signal. This is not to say that global warming is spurious and that we have all been deceived by natural oscillations in the atmosphere; it may well be (and most atmospheric scientists believe this to be so) that the shift in atmospheric flow patterns induced by a recent warming are affecting these natural modes of atmospheric oscillation. Certainly, the NAOI is higher now than it has ever been, and reflects the fact that the six warmest years on record have all occurred since 1990. In fact, 1998 now holds the record for the warmest year since records began (discussed further in the chapter called 'A Personal Perspective' at the end of the book).

The presence of such large, natural variations in temperature in the atmosphere make it difficult to identify the signature of man-made global warming. A crude, first-order method, however, is to simply remove the effects of these oscillations and see what is left. In *plate 5c*, the residual temperature anomalies are shown after the effects of the SO and NAO have been removed. Evidently, there are still large parts of the northern hemisphere which are warmer now than a few decades ago (the red regions), concentrated around the polar regions and straying down into North America, Asia, the tropical and sub-tropical Atlantic, and Africa. But there are also cool regions too, particularly in the Southern Pacific ocean, Mexico, the eastern United States and north-western Europe. If warming is occurring, it is not uniform across the planet. Overall, there does seem to be more warming than cooling, and the net effect should be an increase in global temperatures which, in places, have been as large as 1°C in the past twenty years.

(iv) The Kyoto Protocol

In the first quantified emission limitation and reduction commitment from 2008 to 2012, the assigned amount for each Party ... shall be equal to the percentage ... of its aggregate anthropogenic carbon dioxide equivalent emissions of the greenhouse gases listed in Annex A in 1990... Those Parties for whom land use change and forestry constituted a new source of greenhouse gas emissions in 1990 shall include in their 1990 emissions base year ... the aggregate anthropogenic carbon dioxide equivalent emissions minus removals in 1990 from land use change for the purpose of calculating their assigned amount.

Excerpt from the Kyoto Protocol to the United Nations Framework Convention on Climate Change, December 1997.

In December 1997, the ancient Japanese city of Kyoto played host to an historic meeting, a meeting which takes us a step closer to avoiding at least some of the possible threats to our climate. International negotiations which began in the late 1980s led to a

United Nations Framework Convention on Climate Change in Rio de Janeriro in 1992. The objective of the Convention was to find ways to *stabilise greenhouse gas concentrations at a level that would avoid harmful interference with the climate system.* The Convention's Parties met again in Berlin in 1995 when the decision was made to aim for a specific Protocol at the Third Meeting of the parties in Kyoto, Japan. This Protocol represents an initial step towards setting legally binding limits on greenhouse gas emissions for the first time.

At the time of writing, the Kyoto Protocol remains unratified; it will not come into effect until enough countries have signed it. Whether or not the Protocol will become effective depends on a number of factors, not least among them the inclusion of the world's developing countries. At present they are not included, and yet, as already mentioned, by around 2010 greenhouse emissions from these countries will eclipse those of their developed neighbours. China, with a huge and rapidly growing population and energy generation totally dependent on coal, currently has no plans to sign the Protocol. Its feelings about future restrictions were reflected in a remark made at the Kyoto meeting, indicating that after the west has enjoyed the unrestricted use of fossil fuels for more than a century, the Chinese nation did not intend to enter the twenty-first century on buses and bicycles.

Prior to discussing the specifics of the Protocol itself, it is interesting to look at who was represented at the meeting in Kyoto. Altogether, some 10,000 people attended: five thousand of these were from the industrial sector, some of whom had come to the meeting with technological solutions to the problems under discussion. A further two thousand, five hundred attendees were members of the world press; one thousand represented the various environmental groups; a further 1,000 were local and United Nations support staff; five hundred were government representatives comprising 300 negotiators (although only fifty of these were key personnel) and 200 advisors of all types. Whilst the meeting was intended as an international forum to debate the issues and ramifications of altering the climate, it is perhaps revealing that a mere *twenty* were climate scientists. Many governments chose not to avail themselves of the scientific expertise in their own countries, a curious omission, one might think, when the subject under discussion is very much in the purview of science.

What is the Kyoto Protocol all about? An overview can be

gleaned from figure 25 which provides a time-line of events from 1988 up to the end of the century.

(a)   Tailored Greenhouse Gas Emissions

The Protocol requires that the 39 industrialised countries involved in its formation should limit their annual percentage greenhouse gas emissions between now and the time period 2008–2012, to a specified percentage of their 1990 emissions. However, there are differential responsibilities; not all countries would be required to reduce their emissions of greenhouse gases by the same percentage. In fact, some countries would even be allowed to increase emissions during the next decade. For example, the European Union would be required to reduce emissions by 8%, the U.S.A. by 7% and Japan by 6%; New Zealand would make no change, whilst Australia and Iceland would be permitted to increase their emissions by 8% and 10%, respectively. The net effect for the group of industrialised nations would be a reduction of approximately 5%, compared to emissions in the year 1990. What this actually means is this: if we soon begin making these reductions, then by 2010 the overall emissions should be around 25% less than they would otherwise have been had we pursued a policy of *business-as-usual*.

Obviously, these goals cannot be achieved without the co-operation of industry; it will be necessary to make existing and new technologies accountable so that greenhouse gas emissions can be accurately monitored. Highly policy-relevant questions also arise: is the change in an observed atmospheric trend of a greenhouse gas a response to a Protocol-induced emission reduction, or is it due to a natural change? This eventual accountability is directly analogous to that being addressed currently under the 1987 *Montreal Protocol on Substances that Deplete the Ozone Layer*, with regard to reductions in emissions of ozone-depleting gases.

(b) The Gases Involved

The emissions of six gases would be included in the so-called basket approach: carbon dioxide (CO_2), methane (CH_4), nitrous oxide (N_2O), hydrofluorocarbons (HFCs), perfluorocarbons (PFCs) and sulphur hexafluoride (SF_6). A country's limit would be a radiation-weighted sum for the six; that is, the severity of the effect each gas is believed to have on the radiative balance of the

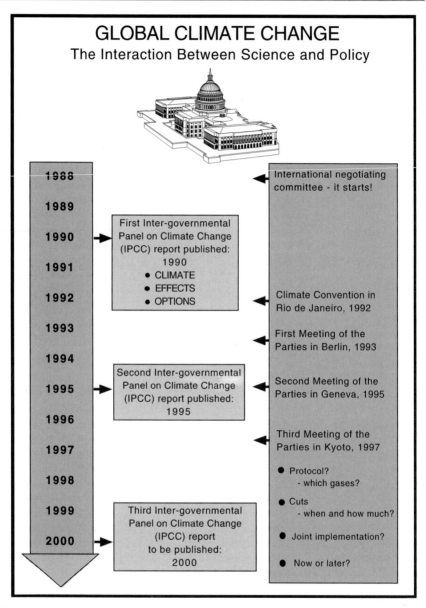

Figure 25 A time-line showing the history of discussions about climate change, beginning in the late 1980s. The Protocol proposed in Kyoto in December 1997 is the first designed to limit future emission of greenhouse gases into the atmosphere. *(Courtesy of Danial Albritton, NOAA Aeronomy Laboratory).*

atmosphere would be weighted accordingly, taking into account such things as the relative abundance of the gas and its individual global warming potential. These factors vary enormously. The release of a single kilogram of SF_6, for example, would introduce (on the time scale of a century) 25,000 times more radiative forcing of the atmosphere than a kilogram of CO_2. Moreover, SF_6 has a half-life of 30,000 years, so that after thirty millennia half a kilogram of SF6 would still be present in the atmosphere. CO_2, on the other hand, has an atmospheric half-life of less than a century.

These, then, are the source gases which the Protocol aims to restrict. On the positive side, *sinks*, where they exist, will also be recognised. A sink, in this context, is a natural reservoir for a substance such as a greenhouse gas which effectively removes it from the atmosphere. For example, growing forests and phytoplankton are responsible for a large uptake of atmospheric carbon dioxide, the amount varying strongly with season. On the negative side, the widespread destruction of forests and the pollution of the oceans has for some time been having a detrimental impact on carbon uptake. Furthermore, biomass burning (the burning of rain forests) to clear land in this century has liberated many millions of tonnes of carbon back into the atmosphere.

The Kyoto Protocol would recognise specific human-influenced increases in carbon storage via forestry (planting trees) as a means of removal, and these efforts will enter negatively into a nation's net emissions, counting as a reduction. Points will be awarded or subtracted for the planting or cutting down of trees, which will contribute to the emissions tally for each country. This action demonstrates a recognition of the scientific fact that the atmospheric abundance of greenhouse gases results from a net sum of emissions (sources) and removals (sinks). Current scientific understanding limits our ability to evaluate most types of greenhouse gas sinks at present, so for the time being only forests are to be included.

(c)   The Emissions Marketplace

During the Kyoto meeting, it was felt by many countries that a trading mechanism would allow global reductions to be made more cost-effective. Experience in the United States with the trading of sulphur dioxide (SO_2) emissions, permitted in the mitigation of acid deposition, is deemed to support this concept. The details of this mechanism are scheduled to be negotiated at the Fourth Meeting of the Parties in 1999. Trading emission quotas is only part of the picture, however; there could also be *emission-partnerships* (also referred to as *bubbles*). The Protocol would recognise that nations could bond together in groups and contribute different levels of emission reductions to achieve the agreed-upon target for overall reductions; the European Union is a standing example. The intention is to provide added flexibility towards lowering net emissions in a more cost-effective and equitable manner. Perhaps the only problem with this approach (in the short term at least) is that some countries are immediately in a position to trade, and therefore have an advantage over those whose emissions are higher.

(d) Developing Countries

As already mentioned, the Kyoto Protocol does not currently place emission-reduction requirements on the countries of the developing world, even though their emissions are set to rise dramatically in the near future, especially China and India. The issue of placing emission-reductions on developing countries will be raised at the next Meeting of the Parties to the Climate Convention.

The problem of climate change is clearly a global phenomenon, and it is to be hoped that the Protocol can promote co-operation between developed and developing countries to tackle this issue. One mechanism for achieving this is the inclusion of a *Clean Development* clause via which a developed country may earn credits from emission-reduction projects in a developing country.

(e) Periodic Reviews of the Kyoto Protocol

Much like the 1987 Montreal Protocol on Substances that Deplete the Ozone Layer, the Kyoto Protocol has a built-in recognition that scientific understanding will evolve and improve our understanding of the mechanisms controlling the climate system. It provides an explicit channel for the scientific community to communicate that improved understanding to decision makers in the world's governments, and such information will be periodically reviewed by the member nations to decide on the need for amendments, if any, in the light of the best available information from updated scientific, technical, social and economic information assessments.

At present, such assessments are updated every five years; the next is expected around 2000/2001.

Perhaps the major stumbling block for the Kyoto Protocol, in its present form, is that contains a clause allowing any member of the 39 developed countries to withdraw from participation four years after it has been ratified. In other words, when the going gets tough, a country has the option to simply back out. If this Protocol is to work, it clearly needs to be tightened-up so that its effectiveness is not subject to the ebb and flow of local economies.

How likely is this to happen? Certainly, the restrictions that such a protocol would impose are sure to impact on economic growth. Our entire concept of energy production is based on carbon, and curbing emission of CO_2 will inevitably restrict production and consumer behaviour; e.g., an increase in the cost of goods produced by industry, and in everyday activities such as driving a car. Ultimately, we may be forced to make a choice between short-term comfort and long-term environmental stability. Living in an unfavourably-perturbed climate, with the attendant changes in weather patterns and increasingly severe storms, could become a very expensive liability in terms of property damage and failed crops.

Of course, as intimated earlier, not everyone believes that climate change is even a real issue. In 1995, a science research authorisation passed by the United States House of Representatives was designed to ban the Environmental Protection Agency (EPA) and the National Oceanic and Atmospheric Administration (NOAA) from conducting long-term studies on climate change. Ironically, this Congressional action came within a few weeks of several reports which advocated an urgent need for just such research to be performed.

The adverse reactions of policy makers and the public alike seems to stem largely from the opinions of those, both within and outside the scientific community, who challenge the scientific basis of climate change. As Dr. George Reid of the NOAA Aeronomy Laboratory in Boulder, Colorado put it in a recent letter to an American newspaper: *Public awareness of the facts is often strongly influenced by minorities on both sides of the issue, and cute phrases are used to catch the public attention.* It is true that the nay-sayers are far less numerous than those who believe that we are on a collision course with a climatic crisis. Alas, they receive a disproportionate amount of publicity, perhaps out of a profound willingness to air their views. According to Ross Gelbspan, author of *The Heat is On: The High Stakes Battle Over Earth's Threatened Climate*, some of this negative activity is being funded by interested industrial parties to create the impression of a divided community. In the scientific investigation of any subject, healthy scepticism is actually encouraged; it is, after all, the very basis of scientific research.

One of the most frequently cited objections to claims that the climate is warming focus on the widely-publicised 0.5–0.6°C increase in temperature over the past century. This rise has been deduced by combining temperature records from many sources, some of which containing large uncertainties. Further doubt is cast by highlighting the discrepancy between ground-based temperature trends and those derived from satellite data, but as mentioned in chapter 3.2, the uncertainties in the latter set of observations are too large for a meaningful comparison. Certainly, when we factor in the large natural variability in the climate, as attested by the coming and going of ice ages, and on shorter scales by fluctuations in atmospheric circulation induced by wave activity, the claim that the world is warming because of human activities does seem somewhat dubious.

But, this being the case, why is it that the vast majority of atmospheric/climate scientists agree that the world really is warming up? To be fair, there is less of a consensus regarding the proportion of this change which may attributed to natural sources, and how much to human activities. Speaking generally, scientists are an innately cautious bunch, anxious to avoid tarnishing their professional reputations. That a consensus exists at all is mainly for one reason: whatever the source of the ground-based temperature data, they all show a change (albeit of different magnitudes) in the same sense, that sense being an increase. Moreover, few people really doubt that carbon dioxide and methane have both increased substantially over the past century, or that these increases now exceed the magnitudes by which they change in response to the passage of ice ages. Which leads to the most important and difficult question of our time: do we make efforts now to reduce later impacts of climate change, assuming it really is happening, or do we wait for climate change to be confirmed to everyone's satisfaction before taking action, thereby committing ourselves to live with the consequences of such a delay? The decision is not going to be easily made because it involves international co-operation, technical feasibility and considerations of equity.

Alas, even if we should decide to act now in the hope of minimising possible future impacts on the climate, there is still the problem of our incomplete understanding of the climate system itself. It is almost certainly foolhardy to attempt changing something when we don't really know how it works! We are still dependent upon observational science and our ability to model the real atmosphere to fill in the blanks.

To wait, or not to wait? That is the question.

As mentioned earlier, the Kyoto Protocol will take effect when or if it is ratified by at least 55 nations, as long as these nations represent 55% of the 1990 carbon dioxide emissions by the 39 industrialised countries. The most promising aspect of the Protocol which may lead to its ratification is its flexibility, as opposed to a *one-size-fits-all* approach. It will be a dynamic document, as indeed is the Montreal Protocol, so that as information is gained by its advisory body, it will be amended accordingly. The IPCC (1996) statement of "a discernible influence on climate" was heard around the world, the report bringing key topics associated with climate change to the fore. The impacts of emission reductions on technology, economics and sociology, as well as the legal, political and equity issues, are far reaching indeed, and they will unquestionably be difficult to resolve. Finally, the time-span over which action must be taken makes it clear that the problem of climate change will span more than a single generation, and will require a multi-generation solution.

Appendices for Part Three

APPENDIX 3.1

GLOBAL WARMING POTENTIAL

The Global Warming Potential, or GWP, is an attempt to provide a simple measure of the radiative effects arising from the emissions of various greenhouse gases. It is an index defined as the cumulative radiative forcing now and at some future time, relative to a reference gas (CO_2). There are problems with this approach, however. It requires a knowledge of where all the emitted gases go, which we currently don't possess. CO_2 is a case in point, where several billion tonnes of emissions every year remain unaccounted for in the carbon cycle. GWP values are usually quoted as single values, but in reality the uncertainty is large, typically 35%.

The net GWPs for the ozone depleting gases, which include direct warming and indirect cooling effects, have been estimated. CFCs, collectively, having a net warming effect, whilst halons (the gases used in fire extinguishers) tend to cool the atmosphere. The global warming potentials of a number of greenhouse gases projected 20, one hundred and 500 years ahead, relative to today's values, are enumerated in Table 8.

Especially noteworthy is the gas HFC-23 which has a truly enormous global warming potential, as has HFC-236fa. All HFC gases are scheduled to be phased out by the year 2029. At the moment, there is no information available as to the nature of their successors, if indeed any are currently planned.

Table 8 Global warming potentials of a number of greenhouse gases. (Adapted from Climate Change 1995: Contribution of Working Group I to the Second Assessment Report of the Intergovernmental Panel on Climate Change).

Species	GWP: 20 years	GWP: 100 years	GWP: 500 years
Carbon dioxide	1	1	1
Methane	56	21	6.5
Nitrous oxide	280	310	170
HFC-23	9,100	11,700	9,800
HFC-32	2,100	650	200
HFC-41	490	150	45
HFC-43-10mee	3,000	1,300	400
HFC-125	4,600	2,800	920
HFC-134	2,900	1,000	310
HFC-134a	3,400	1,300	420
HFC-152a	460	140	42
HFC-143	1,000	300	94
HFC-143a	5,000	3,800	1,400
HFC-on	4,300	2,900	950
HFC-236fa	5,100	6,300	4,700
HFC-245ca	1,800	560	170

APPENDIX 3.2

RADIATIVE FORCING

Radiative forcing is discussed at great length in the Intergovernmental Panel on Climate Change (IPCC) report called 'Radiative Forcing of Climate Change and an Evaluation of the IPCC IS92 Emission Scenarios', published in 1994, and the reader is referred to this publication for an extensive treatment of the subject. The following summarises some of the key issues as they pertain to this book.

A specific definition of radiative forcing was adopted in the IPCC report published in 1990:

The radiative forcing of the surface-troposphere system (due to a change, for example, in greenhouse gas concentration) is the change in net irradiance (in watts per metre-squared; Wm^{-2}) at the tropopause AFTER allowing for stratospheric temperatures to re-adjust to radiative equilibrium, but with surface and tropospheric temperatures held fixed at their unperturbed values.

Paraphrasing from the IPCC 1994 report, the entire concept of radiative forcing actually comes from climate model calculations which suggest an approximately linear relationship between the global-mean radiative forcing at the tropopause and the equilibrium global-mean surface temperature change. Moreover, this relationship seems unaffected by the nature of the forcing, be it a change in solar activity or in greenhouse gas concentrations.

Expressed mathematically, a change in the global-mean radiative forcing may be represented as ΔF (in Wm^{-2}) and the corresponding response in global-mean surface temperature, ΔTS (in K), viz.:

$$\Delta T_S = \Gamma \Delta F$$

where Γ is a parameter which represents the sensitivity of the climate to change which depends on several different processes, including the feedback mechanism associated with water vapour (cloud formation), and the albedo (reflectivity) of polar ice. Its value is hard to determine precisely, but probably lies in the range 0.3–1.4 K/(Wm^{-2}).

Factors affecting radiative forcing are changes in water vapour, ozone and the other greenhouse gases, including HCFCs and HFCs, listed in Table 5 (chapter 3.1). The largest contribution to greenhouse gas radiative forcing made by a *trace* gas is from carbon dioxide: ~1.56 Wm^{-2}, followed by methane at 0.45 Wm^{-2} (although the range of uncertainty here is very large), and nitrous oxide at ~0.15 Wm^{-2}. These values seem small when compared to the forcing of around −4.0 Wm^{-2} induced by volcanic aerosols injected into the atmosphere during major eruptions, as in the case of Mount Pinatubo in 1991, but then the contribution from such an eruption lasts for only a few years, whereas greenhouse gas concentrations are continuously increasing. The above factors are summarised in Figure 26.

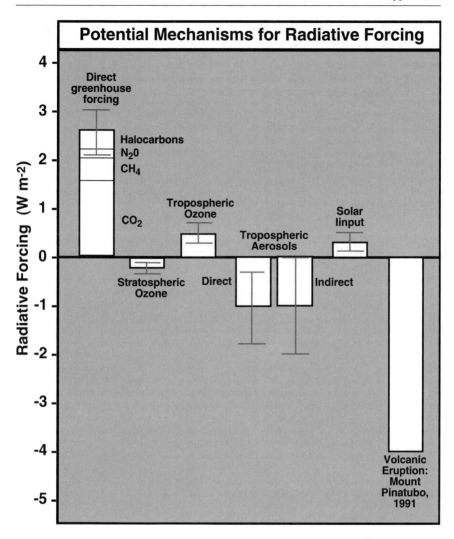

Figure 26 Comparison of the positive radiative forcing of greenhouse gases, including tropospheric ozone, with the negative forcing of aerosols and volcanic eruptions. The eruption of Mount Pinatubo in the Philippines in 1991 produced a forcing of -4 W m^{-2}, but this cooling effect lasted for only a few years. *(Adapted for the IPCC 1994 report)*.

ERRORS IN THE SURFACE TEMPERATURE RECORD

Because satellites have only been used to monitor global temperatures for a few decades, the main evidence for global warming comes from the surface temperature record. This tells us that, during the past century, the mean surface temperature (at least over the world's land masses) has risen by approximately 0.5–0.6°C. Although many sources of error and bias have been identified and corrected in the surface record (e.g., publications by Jones and Wigley, 1990, Jones and Briffa, 1992 and Jones 1994; detailed in the Bibliography for Part Three). Jones et alia (1998) gave recently-generated confidence intervals for the mean annual temperatures since the year 1951 of ±0.12°C, and ±0.18°C prior to 1900. Considering a *decadal* rather than an annual average, the uncertainty is only ±0.10°C for 1951–1995, and 0.16°C for 1851–1900.

Not every one agrees that the errors have been adequately corrected. Daly (1998) believes that many uncorrected errors are positive biases, adding to the observed upward trend. He suggests that the above uncertainties in the temperature record are substantially larger than quoted by Jones and co-workers, the main sources of error believed to be as follows:

(i) Urbanisation around many measurement stations affects the temperature of the air locally — roads and buildings absorb more energy than soil and plants.

(ii) Many different thermometer scales and methods of calibration have been used since this device was invented two hundred years ago, although these were standardised into the Fahrenheit (F) and Celsius (C) scales around 1860. The record which shows a warming surface temperature starts from this time.

(iii) Most thermometers were (and indeed still are) graduated in whole degrees, which means the smallest temperature measured throughout most of its 140-year-long history was 1°F (~0.6°C), although results were usually quoted to a tenth of a degree. The human mind can certainly interpret measurements to this kind of accuracy without the aid of a graduation on the thermometer glass, but it does allow a measure of subjectivity. Towards the end of the nineteenth century, it was also realised that glass shrinks with age, an effect which causes a spurious upward drift in temperature.

(iv) Thermometers do not measure just air temperature but a combination of air temperature affected by the movement of the air itself, and radiation absorbed from its surroundings. In 1884, the Royal Meteorological Society recommended the use of a Stevenson screen to create a standard radiation environment for the thermometer. Although the screen was widely adopted, not all of them were painted white!

(v) At the start of the currently-accepted surface temperature record, in 1860, only 5% of the northern hemisphere was being sampled and the data reported. This rose to 10% by 1880, but was as low as 2% in the southern hemisphere. By 1960, the coverage in the two hemispheres had risen to 50% and 20%, respectively, whereafter these declined to around 35% and 12%. Some allowances are made for the problems introduced by these sampling biases, such as using records from individual sites only, but the correction can hardly be perfect and some uncertainty remains.

(vi) The discrepancy between the surface temperature record over the past twenty years, and the temperature record obtained by satellites, has already been discussed. In addition to orbital decay corrections and other technical problems, there are two factors to consider: (a) thermometers measure air temperatures directly, whereas satellite-borne instruments are, by virtue of their remote locations, *remote sensing devices*; they do not actually measure the same thing as a thermometer, and (b) satellites sample the entire atmosphere which the surface temperature record does not — around 70% of the planet is covered by water.

These various errors and biases almost certainly contribute to changes in the surface temperature record, although the uncertainties are thought by most to be smaller than the observed temperature change by a factor of 3 or 4. The issue is unlikely to be resolved in the near future, and may remain open until a much longer time-series of high quality measurements is available to either confirm or refute global warming.

BIBLIOGRAPHY FOR PART THREE

As with the bibliography at the end of Part Two, I have listed both books and scientific articles, keeping as far as possible to those which are easily accessible. There are scientific reports on climate change, as with the ozone depletion issue, and the most recent of these are included below.

Books

Graedel, T. J. and Paul J. Crutzen. "Atmospheric Change: An Earth-System Perspective". W. H. Freeman, 1993.
Co-authored by the 1995 Nobel laureate in chemistry, this is an excellent book aimed at the college student. It is heavily illustrated and serves as a good introduction to the physical sciences.
"Climate Change, 1994: Radiative Forcing of Climate Change". Edited by J. T. Houghton, L. G. Meira Filho, J. Bruce, Hoesung Lee, B. A. Callander, E. Haites, N. Harris and K. Maskell. Published for the Intergovernmental Panel on Climate Change by Cambridge University Press, U.K.
This volume concentrates chiefly on how the energy balance in the atmosphere may shift in response to increasing greenhouse gas emissions. It also provides an evaluation of a number of emission scenarios.
"Climate Change, 1995: The Science of Climate Change. Contributions of Working Group 1 to the Second Assessment Report of the Intergovernmental Panel on Climate Change". Published for the Intergovernmental Panel on Climate Change by Cambridge University Press, U.K.
This volume includes an overview of the climate system and discusses modelling that system at some length. There is also a section on the detection of the climate change signal.
Gelbspan, R. "The Heat is On: The High Stakes Battle Over Earth's Threatened Climate". Hard-cover, published by Perseus in 1997. Approximate cost £15.00. ISBN: 0201132958.

Seinfeld, John H., and Spyros N. Pandis. "Atmospheric Chemistry and Physics: Air Pollution to Climate". Published by John Wiley & Sons in 1997. Cost approximately £40.00. ISBN: 0471178160.

Publications in scientific journals and reports

Christie, J. R., R. W. Spencer and E. S. Lobl. "Analysis of the Merging Procedure for the MSU Daily Temperature Time Series". *Journal of Climate*, volume **11**, #8, pp. 2,016–2,041, August 1998.

Convention on International Civil Aviation (ICAO): "International standards and recommended practices: environmental protection", *Volume II of "Aircraft engine emissions". Second edition, July 1993.*
This report provides standards for limiting the emissions of smoke, unburnt hydrocarbons (HC), carbon monoxide (CO) and oxides of nitrogen (NO_x) from turbojet and turbofan aircraft engines.

Daly, J. "What's wrong with the surface record?" can be found on the World Wide Web at http://www.vision.net.au/~daly/surftemp.html

Graf, Hans F., I. Kirchner and J. Perlwitz. "Changing lower stratospheric circulation: The role of ozone and greenhouse gases". *Journal of Geophysical Research*, volume **103**, pp. 11,251–11,261, 1998.

Hansen, J., M. Sato and R. Ruedy. "Radiative forcing and climate response". *Journal of Geophysical Research*, volume **102**, pp. 6,831–6,864, 1997.

Hoerling, M. P. and A. Kumar. "Why do North American climate anomalies differ from one El Niño event to another?". *Geophysical Research Letters*, volume **9**, pp. 1,059–1,062, 1997.

Jensen, E. J. and O. B. Toon. "The potential impact of soot particles from aircraft exhaust on cirrus clouds". *Geophysical Research Letters*, volume **24**, pp. 249–252, 1997.

Jones, P. D., and T. M. L. Wigley. "Global Warming Trends". *Scientific American*, August 1990.

Jones, P. D., and K. D. Briffa. "Global surface air temperature variations during the twentieth century: Part 1, spatial temporal and seasonal details". *The Holocene*, volume **2**, p. 165, 1992.

Jones, P. D. "Hemispheric surface air temperature variations: A re-analysis and an update". *Journal of Climate*, volume **7**, p. 1,794, 1994.

Madronich, S., and F. R. de Gruijl. "Skin Cancer and UV radiation". *Nature*, volume **366**, p. 23, 1993.

Miller, R. L. and A. D. Del Genio. "Tropical cloud feedbacks and natural variability of climate." *Journal of Climate*, volume **7**, pp. 1,388–1,402, 1994.

National Geographic, October 1998. This issue discusses the subject of population in depth.

Nicholson, S. E., "Recent rainfall fluctuations in Africa and their relationship to past conditions". *Holocene*, volume **4**, pp. 121–131, 1994.

Oltmans, S. J. and D. J. Hofmann. "Increases in lower-stratospheric water vapour at a mid-latitude Northern Hemisphere site from 1981–1994". *Nature*, volume **374**, pp. 146–149, 1995.

Parrilla, G., A. Lavin, H. Bryden, M. Garcia and R. Millard, "Rising temperatures in the subtropical North Atlantic Ocean over the past 35 years." *Nature*, volume **369**, pp. 48–51, 1994.

Petry, H., J. Hendricks, M. Mollhoff, E. Lippert, A. Meier, A. Ebel and Sausen, R. "Chemical conversion of subsonic aircraft emissions in the dispersing plume: calculation of effective emission indices". *Journal of Geophysical Research*, volume **103**, pp. 5,759–5,772, 1998.

Ramanthan, V. "Greenhouse effect due to chlorofluorocarbons: Climatic implications". *Science*, volume **190**, pp. 50–52, 1975. *One of the earliest papers discussing the issue of global warming, following on the heels of the paper by Rowland & Molina in 1974 (see bibliography for Part Two). It demonstrates the ability of CFCs to act as greenhouse gases and warm the troposphere.*

Reid, G. C. "Solar forcing of global climate change since the mid-17th century." *Climate Change*, volume **37**, pp. 391–405, 1997.

Sausen, R., B. Feneberg, and M. Ponater. "Climatic impact of aircraft-induced ozone changes". *Geophysical Research Letters*, volume **24**, pp. 1203–1206, 1997.

Shindell, Drew T., David Rind and Patrick Lonergan. "Will climate change worsen the ozone holes?". *Nature*, volume **392**, 1998. *The authors model the future of the ozone layer as climate change causes the stratosphere to cool and enhance the catalytic chlorine chemistry.*

Thomson, D. J., "The seasons, global temperature, and precession". *Science*, volume **268**, pp. 59–68, 1995.

Thompson, L. G., E. Mosley-Thompson, M. Davis, P. N. Lin, T. Yao, M. Dyergerov and J. Dai. "Recent warming: ice core evidence from tropical ice cores with emphasis on central Asia". *Global and Planetary Change*, volume **7**, pp. 145–156, 1993.

Wentz, F. J. and M. Schabel. "Effects of orbital decay on satellite-derived lower-tropospheric temperature trends". *Nature*, volume **394**, 13 August 1998.

Wuebbles, Donald J., Chu-Fend Wei and Kenneth O. Patten. "Effects on stratospheric ozone and temperature during the Maunder Minimum". *Geophysical Research Letters*, volume **25**, #4, pp. 523–526, 1998.

This paper models what probably happened to the ozone layer during the era of the Little Ice Age, which was at its coldest during the 17th and 18th centuries.

A PERSONAL PERSPECTIVE

In this book, I have discussed one environmental crisis which is beyond refute, the thinning of the world's ozone layer, and one which still remains highly contentious, the human impact on the world's climate. In the atmospheric science community, there are researchers who believe we are adversely affecting our climate, those who flatly disagree and those who maintain that there is insufficient evidence to make a determination either way.

It is probably apparent from the tone of the book that I incline towards the first of these views. My own conviction rests upon what I consider to be the overwhelming (albeit circumstantial) evidence that the climate around us is indeed changing. In the introduction, I cautioned the reader against making judgements on the basis of isolated meteorological events, and yet the foundation for my belief that the climate *is* changing may seem to rest upon nothing more substantial than this. Well, perhaps so. However, it is perhaps more important to ask ourselves this question: if the climate *is* changing, whether by nature's hand or ours, can we afford to ignore it?

Sixty-five million years ago, the dinosaurs perished in response to some change in their environment, although determining exactly what that change was, still less its magnitude, is proving surprisingly difficult. Perhaps the shift in climate at that time was not large at all, but something rather subtle. Whatever the mechanism, it was sufficient to bring about the demise of creatures which had prevailed for over 160 million years. There is also uncertainty about the mechanism underlying climate change today, but a great deal of circumstantial evidence suggests that some kind of change is certainly in progress.

I find the message conveyed by the graphs in **plate 6** quite compelling: **plate 6a** shows the monthly-mean temperature for each month of 1998, up to the time of writing. In every single month, it has been a record-breaking year in which the monthly-mean temperature has exceeded the previous record. This follows closely on the heels of 1997 being, until that time, the warmest

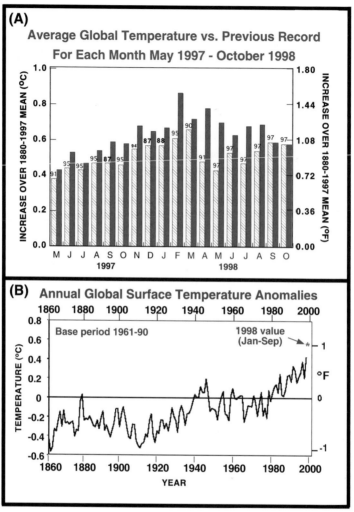

Plate 6 A, In each month between May 1997 and August 1998, the global-mean temperature surpassed the previous record temperature for that month, with the overall value for 1998 more than 0.2°C higher than 1997, itself a record-breaking year; **plate 6B,** temperature anomalies since the beginning of the ground-based thermometer record, showing the decadal variations produced by cycles in solar activity and the sharp positive deviation since the late 1970s. The year 1998 is indicated by a star, and shows by far the biggest temperature increase since the record began. The zero temperature line is defined by averaging temperatures using 1961–1990 as the base period. *(Courtesy of Kevin Trenberth and James Hurrell, National Center for Atmospheric Research).*

calendar year on record. When one reflects that the proposed amount of global warming for this century is of the order of 0.5°C, it is clear that the amount by which 1998 temperatures have exceeded their predecessors (on average, 0.2°C) is by no means trivial. Moreover, the mean temperature for 1997 was already around 0.75°C above those over the last century.

Plate 6b shows the thermometer-derived temperature record since measurements began in 1860. Between 1860 and 1970, the decadal oscillations produced by variations in solar activity can clearly be seen, but subsequent to around 1970 the rise in the mean surface temperature has far exceeded the influence of the Sun's variability. Standing far above all recent temperatures, however, is that for the first two-thirds of 1998. A recent study by Michael Mann and Raymond Bradley (both of the University of Massachusetts), and Malcolm Hughes (University of Arizona) indicates that the last decade of the twentieth century has been the warmest in at least the past 600 years.

Should we then prevaricate and see what happens, or should we act now, just in case these indications should turn out to be significant? It is certainly easy to ignore the evidence when it is not rock-solid: throughout the 1970s and the first half of the 1980s, a seemingly-interminable argument raged over whether or not humans could significantly damage the world's ozone layer. The objections to the idea flew in the face of growing scientific evidence and were often commercially motivated (see the book by Sharon Roan, listed in the bibliography for Part Three, for an in-depth look at the battle between science and industry during this era). Even in the late 1980s, by which time the evidence had become overwhelming, not all the controversy had abated. And still a handful of scientists (if one should dignify them with such a title) insist that the entire business of the ozone hole is a hoax!

The case of ozone depletion provides us with a warning that perhaps we should not wait for totally incontrovertible proof before taking at least *some* mitigating action to safeguard the climate. Had we acted just ten years sooner to restrict or ban CFCs, for example, the ozone layer would recovered decades earlier than currently anticipated. The climate, of course, is a far bigger and more complex system on which we all depend for our survival, and it seems foolhardy to take dangerous risks with it, especially since we still don't properly understand how it works.

Apart from recent extremes in the temperature record, I have reviewed in this book a number of indicators which suggest (to me at least) that a change in climate, man-made or otherwise, is underway. I shall now summarise them before concluding this discussion.

(i)   Glaciers all over the world have been receding throughout the twentieth century, and the polar ice sheets are beginning to melt from beneath, at the interface between ice and sea water, because the sea water has become warmer.

(ii)   Precipitation patterns seem to have altered over much of our planet: the Asian Monsoons have been more sporadic over the past few decades, and when they do come they tend to be more extreme.

(iii) There has been a spate of truly violent hurricanes, such as Hurricane Richard which devastated Nicaragua in October 1998, whilst in many parts of the world, the average amount of rainfall has been lower in the past quarter of a century than at any time since records began.

(iv) Even taking into account the numerous sources of potential error in the land-based temperature record, the magnitude of the recent increases in temperature exceed the worst estimates of those errors.

(v)   Key greenhouse gas concentrations have increased dramatically over the past century, and others (man-made) have been introduced for the first time in the history of the atmosphere. Some of these anthropogenic gases possess enormous global warming potentials, and have lifetimes of thousands of years compared to around a hundred years for carbon dioxide, the most effective naturally-occurring greenhouse gas. This ignores water vapour (but see (vii)).

(vi) Deforestation accounts for up to half of the present rate of growth of carbon dioxide in the atmosphere. By removing the forests, we are effectively diminishing the Earth's ability to absorb CO_2, as well as altering its albedo. Despite localised efforts to save them, the world's rain forests will almost certainly disappear during the twenty-first century, taking with them a substantial fraction of the diversity of living organisms.

(vii) For every 1°C rise in mean surface temperature, it has been calculated that the amount of water vapour in the atmosphere will increase by 6%. This would not only lead to higher levels of precipitation world-wide, but substantially alter the radiative equilibrium of the atmosphere, producing more extremes in weather such as heat waves, droughts, and more frequent and severe hurricanes. The amount of evaporation from the surface of the oceans would be small compared to the thermal expansion of sea water itself, probably flooding large areas of low-lying land around the world.

(viii) Levels of ultra-violet radiation at the Earth's surface are increasing, albeit slowly at the moment. As the ozone layer continues to thin globally at a rate of ~5% per decade, at least for several more decades, UV levels will continue to rise. We are uncertain at which point increased UV light will become deleterious to the ecosystem.

The above list is by no means exhaustive, but it is sufficient for my purposes. I believe that most of these indicators of change can be directly attributed to human activities. We have done so very much to change our environment, and in all likelihood the consequences will be considerably more complex and far-reaching than we presently imagine. Even if these changes have arisen from natural processes alone, we shall still need to deal with their future impacts on our world. So far, we have taken only a few steps, intellectually or technologically, to accommodate them.

It is my personal belief that most of our problems stem from an escalating human population, although not everyone agrees with this point of view. I subscribe to it in part because studies of isolated populations of mammals which have no natural predators ultimately reach such proportions that they undergo a population collapse. We have stepped some of the way out of the framework of natural selection, but we are still confined to one world, and cannot afford to experiment too much with an ecosystem on which we still depend for our very survival.

The other point about population growth is simple ergonomics. At the turn of the century, the number of human beings on the Earth was just 1.7 billion; today, it is around six billion and increasing at almost a billion people per decade. It is hard to see how our world can support almost twice the present human population in the middle of the twenty-first century, even with dramatic advances in agriculture, especially when we consider that two-thirds of those already alive at the end of the twentieth century are either starving or under-nourished. The bigger our population becomes, the less stable our environment is likely to be as we stretch it ever farther to sustain ourselves.

On purely aesthetic grounds, I am not enamoured by the prospect of a planet swamped by human beings at the expense of the many other forms of life with whom we share the Earth. In fact, we are not really sharing it with them at all: we have already curtailed many evolutionary pathways as, one by one, species of plants and animals have fallen prey to our excessive farming activities, both agricultural and hunting/fishing. Their loss is a tragedy because many of them preceded our arrival on the Earth by millions, in some cases *hundreds* of millions, of years. Quite apart from a moral obligation to preserve the diversity of life for future generations, its rapid diminution as habitats change or disappear is yet another warning that our environment cannot support the way we do business.

Our greatest problem may be this: even if the birth-rate were to flatten out today, the population will still continue to expand for several more generations (even if each couple bears only two children) until it matches the death rate. In the best case scenario, the Earth will still need to sustain something like eight billion human beings, and as I indicated above, it is presently having difficulty coping with six billion.

Many countries where populations are rising sharply are only now acquiring the technologies which will enable them to affect the environment significantly, as we in the technologically-developed countries have been doing for more than a century. We are hardly in a position to dictate to the peoples of these countries, telling them that they may not enjoy the same luxuries to which we ourselves have become accustomed. The ultimate irony is that if continued global warming destabilises the environment, then more crops are likely to fail, yielding less food to meet the growing demand instead of more.

I see no way to prevent the fact that two thirds of the world is about to go on-line, as it were. Technological advances should help somewhat, but poorer countries will need considerable financial help if they are to take positive action for the future. Sadly, the process of education is slow and will probably take a generation or more to become truly effective, if it ever does. To achieve any kind of radical change, the long-standing respectability of producing large families will need to change also, and I doubt that thousands of years of social customs be can can be so easily swept aside in favour of solving a problem which seems, at best, intangible, especially when viewed against the harsh realities of day-to-day survival.

I believe, in truth, that we have already passed the point of no return and that we are on a collision course with an environmental crisis. I visualise our situation as riding along in a car which seems to be gathering speed at the top of a steep incline, and we have only recently begun to suspect that there may be something wrong with the brakes.

Only time will tell.

GLOSSARY

SAGE	Stratospheric Aerosol and Gas Experiment
SAM	Stratospheric Aerosol Measurement
SOI	Southern Oscillation Index
SST	Sea Surface Temperature
TOMS	Total Ozone Mapping Spectrometer
UNEP	United Nations Environment Program
WCDP	World Climate Data Programme
WOCE	World Ocean Circulation Experiment
WODC	World Ozone Data Center (Canada)
WMO	World Meteorological Organisation

Commonly Used Quantities and Their Units

Force	newton	N $kg\ m\ s^{-2}$
Pressure	pascal	Pa $kg\ m^{-1}\ s^{-2} \equiv N\ m^{-2}$
Frequency	hertz	Hz s^{-1} (cycles per second)
Energy	joule	J $kg\ m^2\ s^{-2}$
Power	watt	W $kg\ m^2\ s^{-3} \equiv J\ s^{-1}$
Thermodynamic Temperature	kelvin	K $\quad$ ($0K \equiv 273.15°C$)
Mixing ratio	ppmv	parts per million (10^{-6}) by volume
	ppbv	parts per billion (10^{-9}) by volume

Chemical Symbols

Br	bromine
Br_2	molecular bromine
BrO	bromine monoxide
CCl_4	carbon tetrachloride
CH_3Br	methyl bromide
CH_3CCl_3	methyl chloroform
CH_3Cl	methyl chloride
CH_4	methane
Cl	atomic chlorine
Cl_2	molecular chlorine
ClO	chlorine monoxide
CO	carbon monoxide
CO_2	carbon dioxide
H_2O	water
H_2O_2	hydrogen peroxide
HOCl	hypochlorous acid
HNO_3	nitric acid

N	atomic nitrogen
NO	nitric oxide
N_2	molecular nitrogen
N_2O	nitrous oxide
NO_2	nitrogen dioxide
O	atomic oxygen
O_2	molecular oxygen
O_3	ozone
OH	hydroxyl

Definitions

Absolute vorticity

A quantity derived from calculating the wind shear associated with atmospheric motion, combined with the Coriolis parameter.

Adiabatic

The motion of a gas performed in the absence of an exchange of energy with its surroundings.

Aerosols

Airborne particles, the best known of which are used in *aerosol* sprays.

Climate sensitivity

The long-term change in the mean surface temperature of the Earth following a doubling of atmospheric carbon dioxide, the GWP reference gas, or some CO_2 equivalent. In a more general sense, the term is used to represent a change in surface air temperature following a unit change in radiative forcing (in units of $°C/Wm^{-2}$).

Coriolis Effect

An imaginary force which arises when the Earth is treated as a stationary, instead of a rotating, body. It acts radially outward from the axis of rotation if an air parcel's motion is in the same sense as that rotation, and radially inward if that motion is opposed to it.

Greenhouse gas

A gas that absorbs radiation at specific infra-red wavelengths which are emitted by the Earth's surface and by clouds.

Ice sheet	A glacier more than 50,000 km^2 in area forming a continuous cover over a land surface or resting on a continental shelf.
Isentrope	A single surface of potential temperature
Potential temperature	The temperature an air parcel would possess if it were brought adiabatically from some level in the atmosphere to 1000 hPa (approximately sea level).
Potential vorticity	A highly-derived quantity obtained by combining wind speeds, temperature and absolute vorticity, which acts as a dynamical tracer of an air parcel (analogous to a chemical tracer), for periods of a few days to a few weeks.
Radiative forcing	A measure of the importance of a potential climate change mechanism, radiative forcing is a perturbation of the energy balance in the Earth-climate system (in units of Wm^{-2}).
Stratosphere	Highly-stratified and stable region of the atmosphere extending from an average altitude of 12 up to 50 km above the Earth's surface.
Thermohaline circulation	The large-scale, density-driven circulation in the oceans, powered by differences in temperature and salinity.
Tropopause	A temperature inversion at an approximate altitude of 12 km which separates the troposphere from the stratosphere.
Troposphere	Turbulent, unstable region between the ground and approximately 12 km, containing all the world's weather systems.

ACKNOWLEDGMENTS

I wish to acknowledge the numerous people who have kindly contributed to this book. Whilst most of the material used is widely accepted in the scientific community to be correct, these contributors are in no way responsible for my interpretation of the facts.

Cover images: Ann McCarthy at the NOAA National Severe Weather Laboratory, for the wall cloud and lightning photograph; Dr. Paul Newman at NASA's Goddard Flight Center in Maryland for the original version of the polar ozone depletion plot, and Mr. Jan Bruyndonckx, Belgium, for the photograph of the Sahara desert.

One of the sources I found particularly useful, especially for the summary of the effects of UV radiation on the ecosystem and the ozone issue in general, is the publicly available material written by Dr. Robert Parsons at the University of Colorado. His summaries are accessible at the following World-Wide Web sites:

http://www.faqs.org/faqs/ozone-depletion/
http://www.cis.ohio-state.edu/hypertext/faq/usenet/ozone-depletion/top.html
http://www.lib.ox.ac.uk/internet/news/faq/sci.environment.html
http://www.cs.ruu.nl/wais/html/na-dir/ozone-depletion/.html
Plaintext versions can be found at:
ftp://rtfm.mit.edu/pub/usenet/news.answers/ozone-depletion/
ftp://ftp.uu.net/usenet/news.answers/ozone-depletion/

If you encounter any problems accessing the site, you should send e-mail to Dr. Parsons; his address is: rparson@spot.colorado.edu.

My thanks also to Peter Newton and R. S. Falk, Department of Trade and Industry, London for access to their report *DTI forecast of fuel consumption and emissions from civil aircraft in 2050 based on ANCAT/EC2 1992 data.*

Also, my thanks to Dr. Alan Howells for his contribution to Appendix 2.4.

DATA SOURCES

Ozone data in figures 8A and 15A were used with the permission of Dr. Geraint Vaughan at the University of Wales, Aberystwyth; in figure 9B, of Professor Cristos Zerofos at the Aristotle University of Thessoliniki, Greece; in figure 8B, of the World Ozone Data Center in Canada. Data used for figure 14A was acquired during the European Arctic Stratospheric Ozone Experiment (EASOE) conducted during the winter of 1991-2 (available on a data CD distributed by the Norwegian Institute for Air Research); and for figure 14B during the Airborne Southern Hemisphere Ozone Experiment (ASHOE) conducted it 1994.

In-situ measurements of ozone and chlorine monoxide used for figure 12 were obtained by NASA's high altitude ER-2 aircraft during the Airborne Antarctic Ozone Expedition (AAOE) in 1987.

Meteorological data used to calculate the air parcel trajectories shown in figure 14C,D were provided by the European Centre for Medium-range Weather Forecasting (ECMWF), Reading, U.K.

FIGURES

I wish to express my appreciation to the following people who generously provided graphical material for inclusion in this book.

Dr. Daniel Albritton, Director of the Aeronomy Laboratory, National Oceanic and Atmospheric Administration (NOAA), Boulder, Colorado for the original versions of figures 18 and 25, showing the time-lines for the Montreal and Kyoto Protocols. Dr. Albritton was the Science Adviser to the United States delegation at the Meeting on Climate Change held in the Japanese city of Kyoto in December 1997. These figures are not covered by the copyright restrictions of this work, and anyone interested in using them should seek the permission of Dr. Albritton.

Drs. J. T. Houghton, G. J. Jenkins and J. J. Ephraums, for permission to adapt the figure presenting time series' of temperature, carbon dioxide and methane, which appears in the publication *Climate Change: The IPCC Scientific Assessment: 1991.* My version is presented in figure 20.

Dr. Douglas V. Hoyt, Hughes/STX, U.S.A. for supplying temperature data from 1610 A.D. up to the present.

Dr. James Hurrell, National Center for Atmospheric Research (NCAR), Boulder, Colorado for supplying the original form of **plate 5.**

David S. Lee, AEA Technology plc, Oxfordshire, U.K. for the original version of **plate 4**, which shows the spatial distribution of NO_x emissions from civil aviation for the year 1992, accompanied by those anticipated for the years 2015 and 2050.

Dr. D. Pierce at the Scripps Institution of Oceanography, Experimental Climate Prediction Center, for **plate 3** which shows the heating effect of El Niño in the equatorial Pacific.

NOAA National Climatic Data Center for **plate 6a** showing the record breaking temperatures throughout the first two-thirds of 1998.

Dr. George Reid, NOAA Aeronomy Laboratory, Boulder, Colorado for figure 21 showing sunspot numbers since 1620 A.D.

Dr. Mark Schoeberl, NASA Goddard Space Flight Center, Greenbelt, Maryland for the original version of the upper six images in **plate 1**, showing the development of the ozone hole over Antarctica, 1979–1994, and Dr. Paul Newman, also at NASA/Goddard, for the lower six images showing ozone depletion over the Arctic between 1971 and 1997. Note: this figure is not bound by copyright restrictions, and anyone wishing to make use of this material should contact either Dr. Schoeberl or Dr. Newman at NASA.

Jonathan Shanklin of the British Antarctic Survey for making available the total column ozone data used in figure 11A.

My special thanks to Dr. Drew Shindell at NASA Goddard Institute for Space Studies, New York, not only for providing the original form of figure 17, but also for reading and commenting upon some of the text in this book.

Dr. Richard Stolarski, NASA Goddard Space Flight Center, Greenbelt, Maryland for providing the original form of figure 13, showing mid-latitude ozone trends derived from ozonesonde and SAGE II data.

Drs. Kevin Trenberth and James Hurrell at the National Center for Atmospheric Research, Boulder, Colorado for **plate 6b**, showing annual-mean surface temperatures from 1860 to 1998.

Dr. Joe Waters, NASA Jet Propulsion Laboratory for **plate 2** showing the relationship between temperature, nitric acid, chlorine monoxide and ozone at polar latitudes in both hemispheres. Note: this image is not bound by copyright, and anyone wishing to use this figure should contact Dr. Waters at JPL.

Dr. Donald Wuebbles, Director of the Environmental Council, University of Illinois, for providing the original form of figure 16, showing modelled ozone loss during the Little Ice Age.

INDEX

COLOUR PLATES

Plate 1 Polar ozone loss seen by orbiting satellites. Data from a type of instrument called a Total Ozone Measuring Spectrometer (TOMS), which has been mounted on various satellites since the early 1970s, shows how the ozone column has declined not only over the south pole regions between October 1979 and October 1994, but also in the Arctic where the loss first became noticeable around 1990. Note that the colour scale for the top six images (southern hemisphere) runs from 100 to 450 DU, whereas for the bottom set of six images (northern hemisphere), it runs from 240 to 520 DU. There has always been more ozone in the polar spring than over Antarctica because of the different dynamical behaviour of the atmosphere in each hemisphere. (*Courtesy of Mark Schoeberl and Paul Newman, NASA*).

Plate 2 Satellites are now able to measure numerous chemical species in the atmosphere, and provide additional evidence for the human impact on the ozone layer. As well as temperature, the column amounts of nitric acid (HNO_3, an important intermediary in the ozone destruction process), chlorine monoxide (ClO) and ozone (O_3) are recorded simultaneously in both the northern (top four images) and southern (bottom four images) hemispheres. Notice how areas of low temperature, HNO_3 and O_3 correspond closely to the area occupied by high concentrations of ClO. (*Courtesy of Joe Waters, NASA JPL*).

Plate 3 (a) El Niño is triggered by a slackening of the easterly (from the east) winds in the equatorial Pacific, decreasing upwelling and increasing the surface level on the eastern side of the ocean, and correspondingly increasing upwelling and decreasing the surface level on the western side; (b) the weakened easterlies (winds from the east) cool the eastern ocean surface less and allow the waters to warm up, whilst in the west the waters tend to cool. This creates a strong temperature gradient allowing the surface waters to plunge westward. This movement alters the path of the subtropical jet stream, and this change in the weather pattern is communicated to other regions of the atm osphere. (*Courtesy of D. Pierce, Scripps Institution of Oceanography*).

Plate 4 Spatial distribution of 1992 NO_x emissions (A) from civil aviation, vertically integrated between the ground and an altitude of 16 km, in units of kilograms of NO_2 per metre-squared per year (kg NO_2 m^{-2} yr^{-1}); modelled spatial NO_x distributions for the years (B) 2015 and (C) 2050, compared to 1992. Note the changing colour scale to the right of each figure, on going from A to C. (*Courtesy David S. Lee, A.E.A. Technology plc, Oxfordshire*).

Plate 5   Persistent anomalies in surface temperature (A) and sea level pressure (B) caused by shifts in atmospheric circulation. The residual temperature anomaly (C) cannot be explained by this natural mechanism. Persistent changes are gauged by comparing temperatures at widely-separate sites in the southern (D) and northern (E) hemispheres, a method to which scientists refer as an index (it is actually a pressure index). The red regions in (D) and (E) suggest a persistent warming (compared to earlier years) which has been in progress throughout the last quarter of the 20th century. (Courtesy of James Hurrell, National Center for Atmospheric Research).

Plate 6A   In each month between May 1997 and August 1998, the global-mean temperature surpassed the previous record temperature for that month, with the overall value for 1998 more than 0.2°C higher than 1997, itself a record-breaking year; plate 6B, temperature anomalies since the beginning of the ground-based thermometer record, showing the decadal variations produced by cycles in solar activity and the sharp positive deviation since the late 1970s. The year 1998 is indicated by a star, and shows by far the biggest temperature increase since the record began. The zero temperature line is defined by averaging temperatures using 1961–1990 as the base period. *(Courtesy of Kevin Trenberth and James Hurrell, National Centre for Atmospheric Research)*.

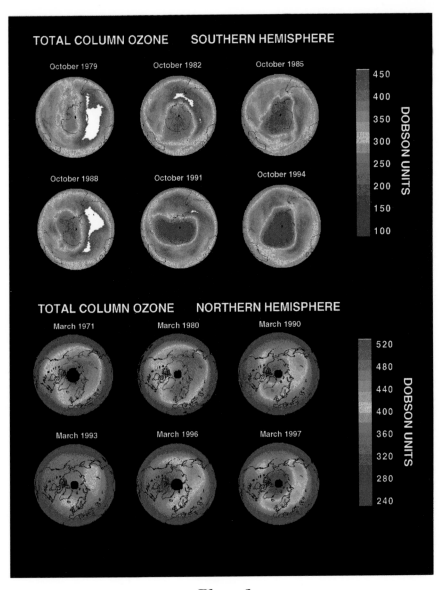

Plate 1

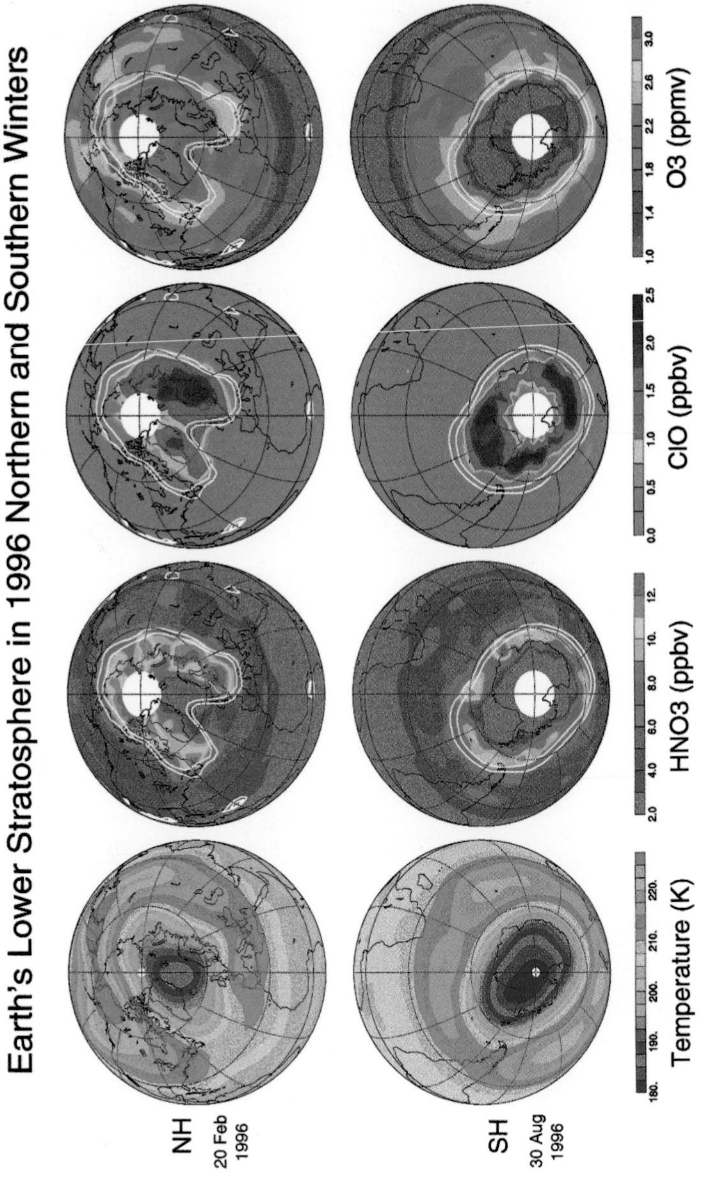
Earth's Lower Stratosphere in 1996 Northern and Southern Winters
NH
20 Feb
1996
SH
30 Aug
1996
180. 190. 200. 210. 220.
Temperature (K)
2.0 4.0 6.0 8.0 10. 12.
HNO3 (ppbv)
0.0 0.5 1.0 1.5 2.0 2.5
ClO (ppbv)
1.0 1.4 1.8 2.2 2.6 3.0
O3 (ppmv)

Plate 2

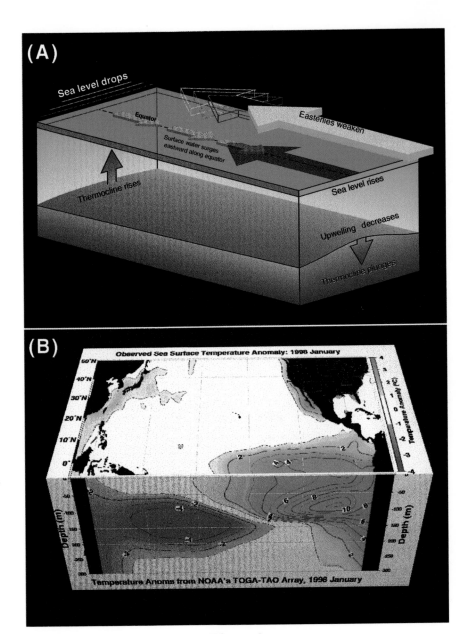

Plate 3

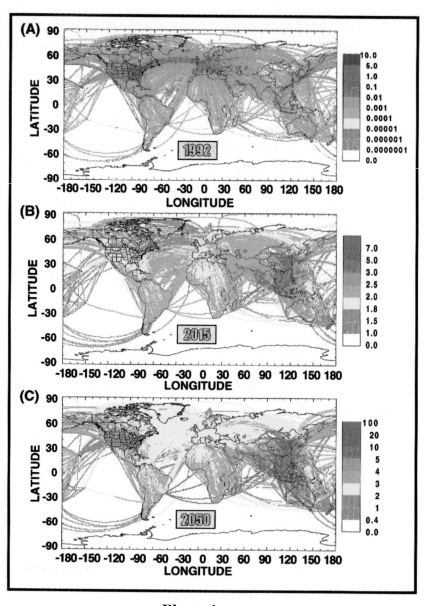

Plate 4

Sea Surface Temperature Anomalies and Temperature Oscillation Indices

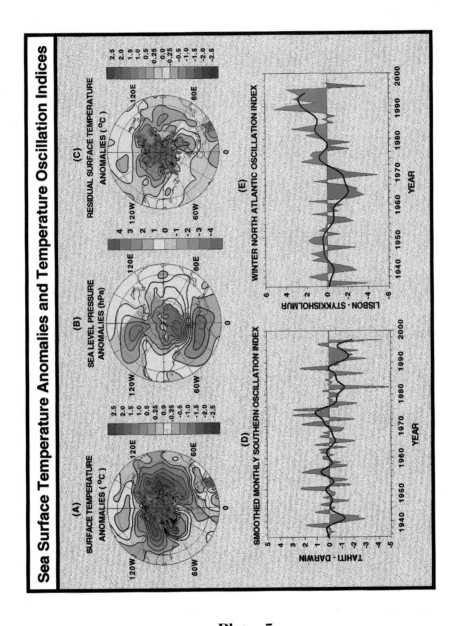

Plate 5

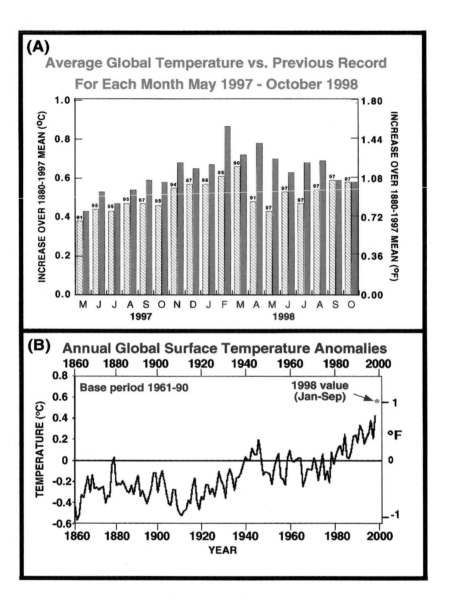

Plate 6